ABOUT THIS REPORT

This report is a compilation of published material covering technology developments that relate directly to spatial technologies or that operate in support of spatial technologies. The intended audience is anyone with an interest in the progressive development of spatial technologies.

ACCESS AND AVAILABILITY

The report is available in PDF format at www.crcsi.com.au
We welcome your comments regarding the readability and usefulness of this report.
To provide feedback, please contact us at info@crcsi.com.au

CITING THIS REPORT

Coppa, I., Woodgate, P. W., and Mohamed-Ghouse Z.S. (2016), 'Global Outlook 2016: Spatial Information Industry'. Published by the Australia and New Zealand Cooperative Research Centre for Spatial Information.

ISBN 978-0-9943019-7-0

COPYRIGHT

DISCLAIMER

While every effort has been made to ensure its accuracy, the CRCSI does not offer any express or implied warranties or representations as to the accuracy or completeness of the information contained herein. The CRCSI and its employees and agents accept no liability in negligence for the information (or the use of such information) provided in this report.

TABLE OF CONTENTS

INDEX OF FIGURES

INDEX OF FIGURES (CONT)

ACRONYMS

ACS	Analysis Centre Software
AI	Artificial Intelligence
AMD	Age-related Macular Degeneration
APP	Software application for mobile devices
APS	Australian Public Services
AR	Augmented Reality
ATAP	Advanced Technology and Projects
BIM	Building Information Modeling
BLE	Bluetooth Low Energy
CAD	Computer Aided Design
CAGR	Compound Annual Growth Rate
CEDA	Committee of Economic Development of Australia
CRCSI	Australia and New Zealand Cooperative Research Centre for Spatial Information
DARPA	Defence Advanced Research Projects Agency
DOD	United States Department of Defense
EO	Earth Observation
GNSS	Global Navigation Satellite System
GPS	Global Positioning System
IDC	International Data Corporation
HUD	Head-up Display
IOT	Internet of Things
ISS	International Space Station
LBS	Location Based Services
OGC	Open Geospatial Consortium
PNT	Positioning, Navigation and Timing
PPP-RTK	Precise Point Positioning- Real Time Kinematic
UAV	Unmanned Aerial Vehicle
UCS	Union of Concerned Scientists
RDF	Resource Description File
RFI	Radio Frequency Interference
RPAS	Remotely Piloted Aircraft System
SDI	Spatial Data Infrastructure
STOIC	Spatial, Temporal and Orientation Information in Contested Environments Program
UML	Unified Modeling Language
UN-GGIM	United Nations Initiative on Global Geospatial Information Management
VR	Virtual Reality
VSAWB	Virtual Spatial Analytics WorkBench
VTOL	Vertical Take- off and Landing Aircraft
WMD	Weapons of Mass Destruction
WMS	Web Map Services

KEY POINTS

- The spatial enablement of the connected devices that make up the Internet of Things is still in its very early stages. It is anticipated that 50+ billion devices will be connected by 2020 with estimates that this market will be worth US$19 trillion within the next 10 years (Cisco).
- Nearly half (3.4 billion people) of the world's population are connected to the internet, and a second 'space race' is currently underway to connect the other half: internet service providers are developing innovative means to complement existing ground-based and traditional satellites including low-earth orbit lightweight satellites, stratospheric balloons and solar drones.
- There will be a rapid increase in the number of active satellites and payloads due to the miniaturisation of space technologies. As of 30 June 2016 there were 1,419 active satellites in orbit with thousands more now under active consideration – for example, OneWeb plans to put 648 communications microsatellites into orbit. Additionally, Samsung has proposed a fleet of 4600 communication satellites in a low-earth orbit.
- As of 30 June 2016, there are 364 operational earth observation satellites in orbit. Noticeable is a shift away from government-funded platforms towards microsatellites that are industry-funded and operated. These platforms often utilise off-the-shelf components and cost a fraction of the traditional satellites. This trend will lead to the availability of cheap and abundant satellite imagery over the next few years.
- Six different Global and Regional Navigation Satellite Systems are operational or under development. On 30 June 2016, there were 105 GNSS satellites in orbit. Ground-based positioning and navigation technologies are also advancing rapidly.
- Location-Based Services (LBS) have experienced explosive growth over the last couple of years. LBS were predicted to grow at a Compound Annual Growth Rate (CAGR) of 54%, from US$1.66 billion in 2013 to US$14.8 billion in 2018. A number of companies are building seamlessly integrated indoor and outdoor location and navigation capabilities for the market.
- The commercial remotely piloted aircraft market (RPAS, also known as UAVs or drones) is experiencing explosive growth with one estimate at 51% CAGR for the period 2014–2019. There are predictions that the market for small RPAS is set to surpass US$8.4 billion by 2018. RPA laws have been changing in various nations, but progress has been more cautious and slower than desired by the market.
- Many governments systematically work on strengthening and extending Open-Data policies. In Australia, several jurisdictions have reported substantial increases in geospatial data requests via their open Globe platforms.

GLOBAL MEGA TRENDS

PricewaterhouseCoopers [1] nominate the five global mega trends as:

- Demographic and social change
- Shifting economic power
- Rapid urbanisation
- Climate change and resource scarcity
- Technological breakthroughs

Frost and Sullivan (2014) identify the mega trends shown in Figure 1, giving the degree of certainty as well as the degree of impact on the global economy.

FIGURE 1: GLOBAL MEGA TREND MATRIX 2025

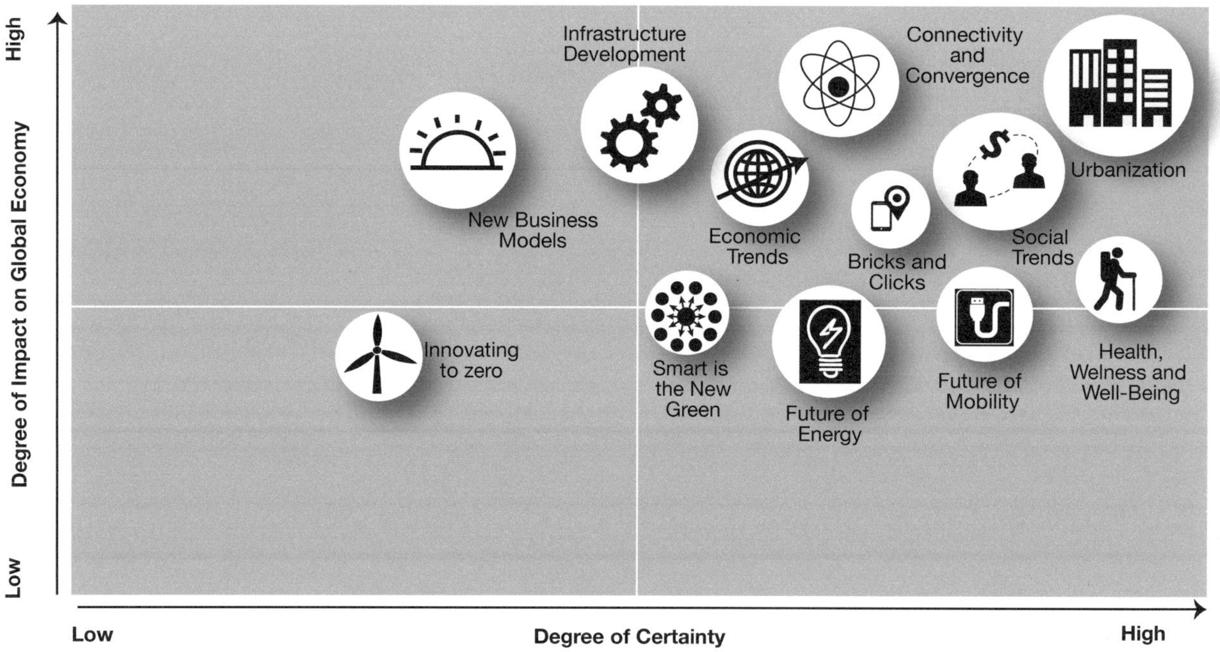

Note: The size of the bubble represents the scale of opportunity within each Mega Trend.
These Mega Trends have been plotted based on quantative and qualitative reasoning.

Source: [2]

Figure 2, also compiled by Frost and Sullivan [2], shows areas of game-changing technologies:

FIGURE 2: TOP TEN TRANSFORMATIONAL SHIFTS BY 2020

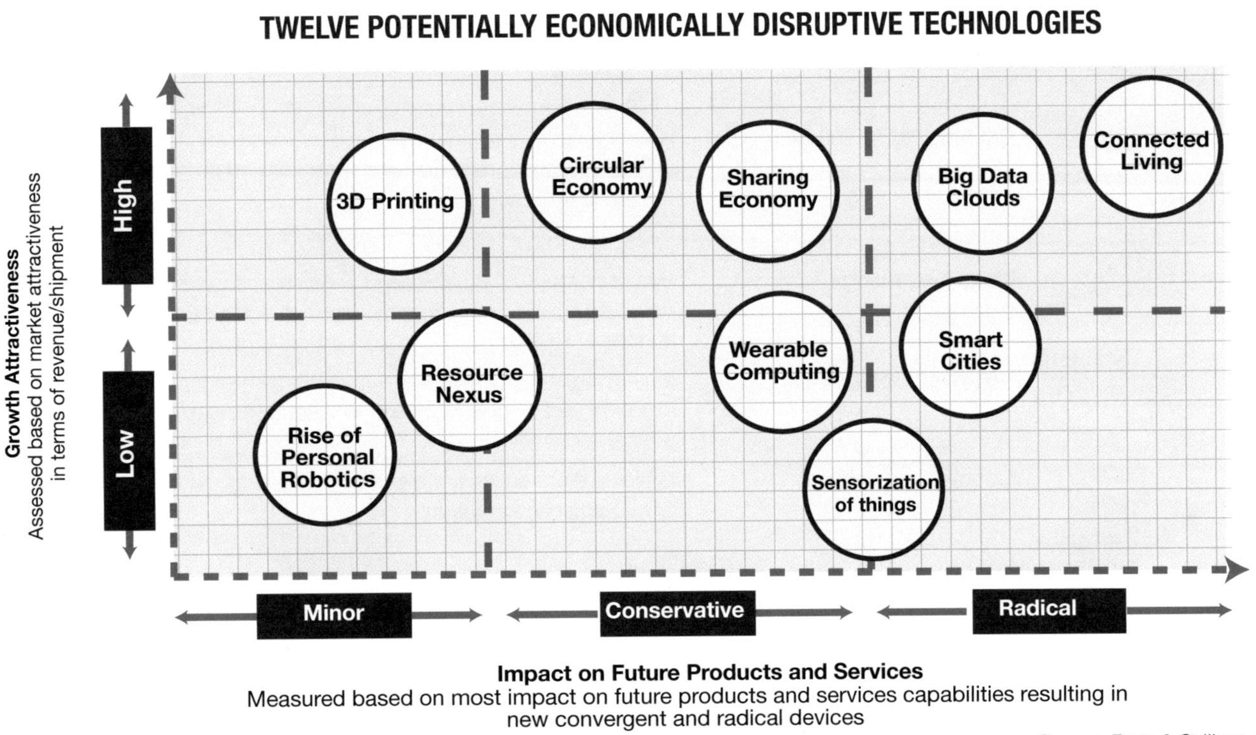

Source: Frost & Sullivan analysis

Additionally, the McKinsey Global Institute has published the following list (Figure 3) of potentially disruptive technologies [3].

FIGURE 3: TWELVE POTENTIALLY ECONOMICALLY DISRUPTIVE TECHNOLOGIES

Mobile Internet	Increasingly inexpensive and capable mobile computing devices and internet connectivity.
Automation of knowledge work	Intelligent software systems that can perform knowledge work tasks involving unstructured commands and subtle judgements.
The Internet of Things	Networks of low-cost sensors and actuators for data collection, monitoring, decision makeing and process optimisation.
Cloud Technology	Use of computer hardware and software resources delivered over a network or the internet, often as a service.
Advanced Robitics	Increasinglly capable robots with enhanced senses, dexterity and intelligence used to automate tasks or augment humans.
Autonomous vehicles	Vehicles that can navigate and ooperate with reduced or no human intervention.
Next-generation geonomics	Fast, low cost gene sequencing, advanced big data analytics and synthetic biology ("writting" DNA).
Energy storage	Devices or systems that store energy for latter use, including batteries.
3D printing	Additive manufacturing techniques to create objects by printing layers of material based on digital models.
Advanced materials	Materials designed to have superior characteristics (e.g. strength, weight, conductivity) or functionality.
Advanced oil and gas exploration and recovery	Exploration and recovery techniques that make extraction of unconventional oil and gas economical.
Renewable energy	Generation of electricity from renewable sources with reduced harmful climate impact.

Source: [3]

The 'disruptive technologies' listed have the potential to rapidly revolutionise the way we live and work; such as the internet and smartphones. Disruptive technologies are well-known for rendering well-established companies and even industries obsolete.

Jeremy Rifkin, President of the Foundation on Economic Trends, states:

> What we are beginning to see are the outlines of a Third Industrial Revolution, a new convergence between communication, energy, and transportation logistics. The communications internet is morphing and converging with an incipient renewable energy internet and a fledgling automated, GPS-guided, and soon driverless transportation internet to create the three internets in one seamless kernel which is atop a platform called the Internet of Things. [4]

It is interesting to note how the pace of disruption has increased exponentially [5]. While it took the telephone 75 years to reach 100 million users worldwide (starting in 1878), it took only two years and four months for Instagram to reach an audience of the same size. In July 2016 the location-enabled augmented reality game 'Pokémon Go' took a mere 25 days to be downloaded by 100 million users [6]. We live in times when the adoption of new technology is occurring at an unprecedented pace.

FIGURE 4: EXPONENTIAL PACE OF DISRUPTIVE TECHNOLOGIES

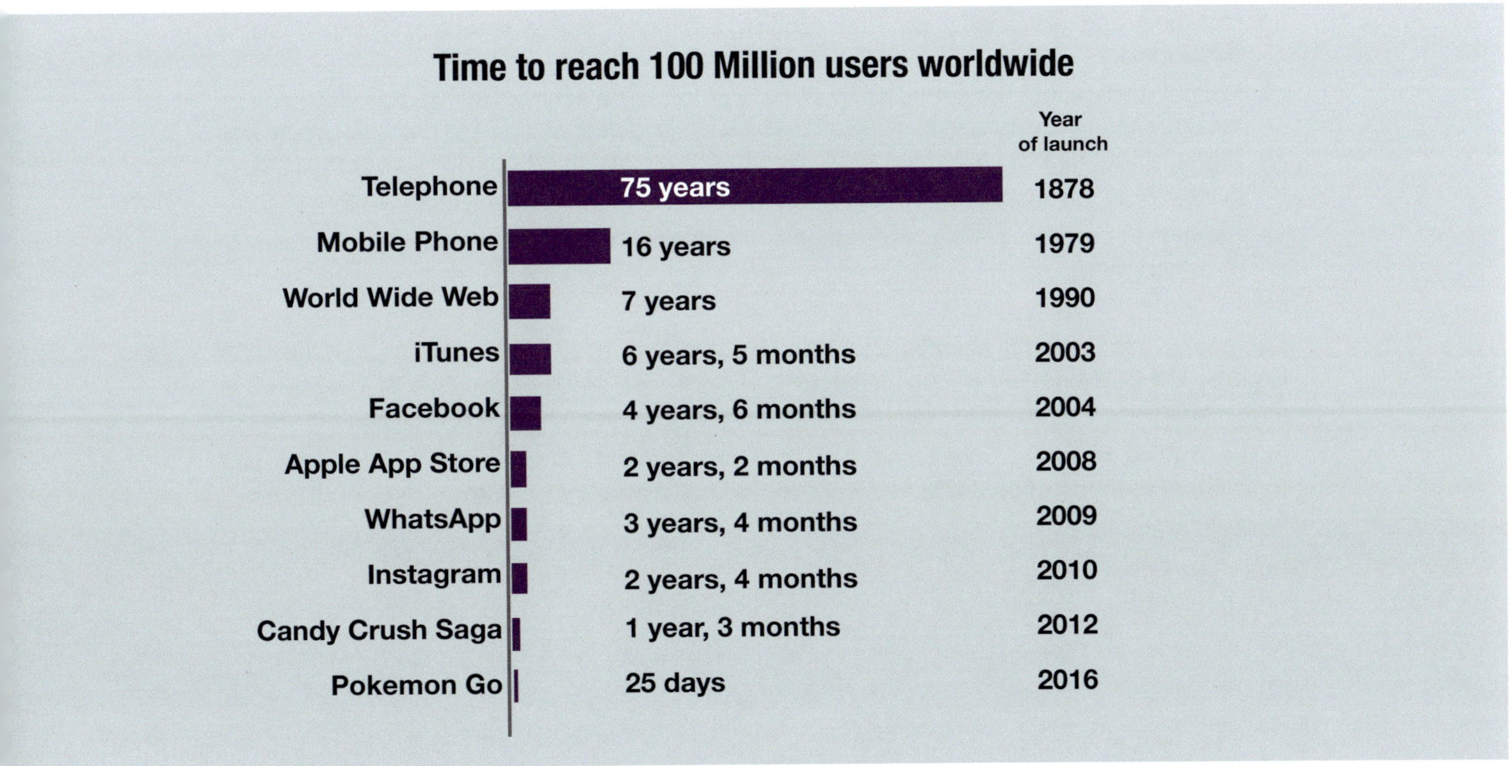

Source: [5] Sources: ITU:Statistia; BCG research. Telephone: ITU; mobile phone: ITU (the base year is considered the start of the first cellular network). Mobilephonehistory.co.uk; World Wide Web: Scientific American, Internet Live Stats; Itunes:number of accounts, Fortune, Facebook: active monthly users, Facebook. Apple App Store: Number of accounts, OSX Daily, VentureBeat; WhatsApp:Active users, Wired, Digital Quarterly; Instagram: monthly users, TechCrunch; Candy Crush Saga: Facebook users only, AppMtr.com

In September 2015, 193 members of the United Nations (UN) unanimously adopted the bold global 'Agenda 2030' for sustainable development.

THE 17 NEW SUSTAINABLE DEVELOPMENT GOALS INCLUDED IN THE AGENDA ARE:

1. End poverty in all its forms everywhere
2. End hunger, achieve food security and improved nutrition, and promote sustainable agriculture
3. Ensure healthy lives and promote well-being for all at all ages
4. Ensure inclusive and equitable quality education and promote lifelong learning opportunities for all
5. Achieve gender equality and empower all women and girls
6. Ensure availability and sustainable management of water and sanitation for all
7. Ensure access to affordable, reliable, sustainable and modern energy for all
8. Promote sustained, inclusive and sustainable economic growth, full and productive employment and decent work for all
9. Build resilient infrastructure, promote inclusive and sustainable industrialisation and foster innovation
10. Reduce inequality within and among countries
11. Make cities and human settlements inclusive, safe, resilient and sustainable
12. Ensure sustainable consumption and production patterns
13. Take urgent action to combat climate change and its impacts
14. Conserve and sustainably use the oceans, seas and marine resources for sustainable development
15. Protect, restore and promote sustainable use of terrestrial ecosystems, sustainably manage forests, combat desertification, and halt and reverse land degradation and halt biodiversity loss
16. Promote peaceful and inclusive societies for sustainable development, provide access to justice for all and build effective, accountable and inclusive institutions at all levels
17. Strengthen the means of implementation and revitalise the global partnership for sustainable development [8]

Geographic information is seen as critical in implementing these goals. The UN Committee of Experts, UN-GGIM (United Nations Initiative on Global Geospatial Information Management) is using spatial information to develop measurable sustainable development goal indicators [9].

In the context of global trends and the rapid development of technologies, disruptive and transformative issues of particular relevance to the spatial industry include:

- The massive network of connected devices and sensors powered by smart applications (also known as the Internet of Things)
- The Space Race to connect more than 3 billion people to the internet
- Wearable technology
- Seamless indoor to outdoor navigation and location enablement
- Imagery everywhere from satellite remote sensing, airborne systems, ground-based cameras, videos and other remote sensing and close sensing imagery devices

SPATIAL TECHNOLOGY TRENDS

INTRODUCTION

This paper builds on previous work published in 2014 [10], and aims to give an update on the developments in various fields that are relevant for spatial technologies (see Figure 5).[1]

FIGURE 5: OVERVIEW TOPICS DISCUSSED IN THIS PAPER

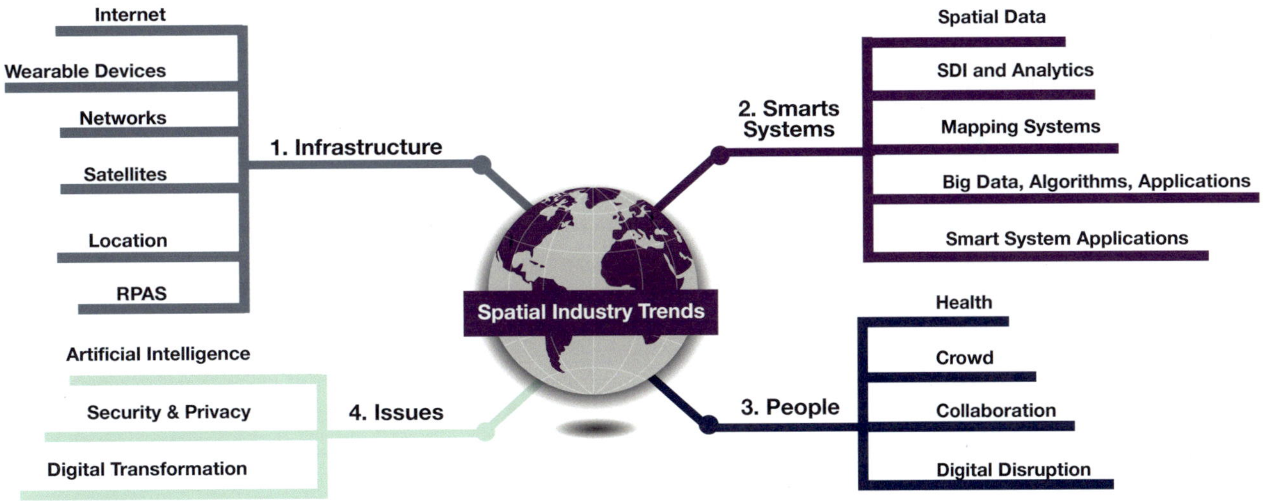

Overall, this part of the paper provides an update on specific developments in the spatial industry. The first section deals with the areas of infrastructure that collect spatial data, including sensors (Internet of Things, wearable and smart devices), networks and connectivity, and remote sensing technologies (satellites and RPAS). Section 2 gives an update on the prerequisites for making systems smart; available open data, spatial information systems, big data, algorithms, applications and examples of the various resulting smart systems.

Section 3 looks at human interaction with spatial systems, from crowdsourcing data and intelligence to the spatial workforce and the relevant technologies touching on human health. In section 4 concerns are discussed, including the development of artificial intelligence and its implications, the complicated conflict between the need for security and the right to personal privacy, as well as the digital transformation agenda.

Technological trends and advances that have a high likelihood of converging with the spatial industry are considered throughout this paper.

[1] The reader is encouraged to review the paper from 2014 to gain a more complete picture.

1. INFRASTRUCTURE

1.1 THE INTERNET

The internet connects devices, people, processes and data in an integrated global network. Cisco estimates this market will be worth US$19 trillion within the next 10 years (with US$14 trillion from the private sector and US$5 trillion from the public sector), created by initiatives such as smart cities and infrastructure [11].

It is anticipated that by 2020 the internet will have 50+ billion connected devices, which in turn will each have multiple sensors, thus creating a trillion sensor ecosystems – also known as the 'Internet of Things' [11]. In effect, it will be possible to build a 'Planetary Nervous System' that will assist individual decision-making and enable self-organising 'smart' systems to create a collective intelligence [12]. To achieve an economy of scale and interworking, interoperable global standards will be very important [13] as will semantically enabled algorithms and applications that streamline standards.

Several companies aim to become standard platforms for the Internet of Things – the spectrum spans from companies such as AGT [14] (offering an open, modular platform with Internet of Things specific analytics tools, data management and visualisation capabilities) to the smart tool ecosystem of the 'SmartThings Hub' [15], recently acquired by Samsung [16].

There is no dominant Internet of Things platform as yet, and several market leaders will probably emerge, seeking to establish international standards and thus dominant interoperability. Established technology companies, such as Cisco, GE and IBM as well as specialised Internet of Things companies such as Jasper, Sierra and PTC, understand the Internet of Things market very well. Their challenge will be to connect developers with end users. Google and Apple have built ecosystems with 5+ million developers combined. In the Q1 developer survey 4000+ developers indicated that Apple and Google's nascent platforms are the most popular; Internet of Things would be an extension of their current activities, and thus Google and Apple are in pole positions for the upcoming Internet of Things platform race [17].

Adelaide-based company Myriota [18] is commercialising technologies developed by the University of South Australia that will use low-earth orbit satellites for two-way data connectivity for remote sensors and devices. It builds on miniaturisation and low-power operations, and expects to bring current costs down by factor 10 [19].

Microsoft commissioned research that found that a full 48% of Australian businesses are either unaware or have no plans to deploy Internet of Things technology. However, the organisations that have implemented Internet of Things and are measuring its effectiveness, report an average of 28% in cost savings, and a 29% productivity and efficiency gain [20]. Around 38% report that revenues have risen (by an average of 33%) after deploying Internet of Things; an increase they put down to improved and accelerated decision-making and better customer service [20].

Figure 6 shows the Hype cycle for a variety of Internet of Things applications, from the innovation trigger to the plateau of productivity stages. As can be seen, most Internet of Things applications are still at an early stage and many are not yet fully commercially implemented. However, the internet of everything is expected to develop rapidly over the next few years, and the adoption rate for this disruptive technology might surprise many.

FIGURE 6: HYPE CYCLE FOR THE INTERNET OF THINGS

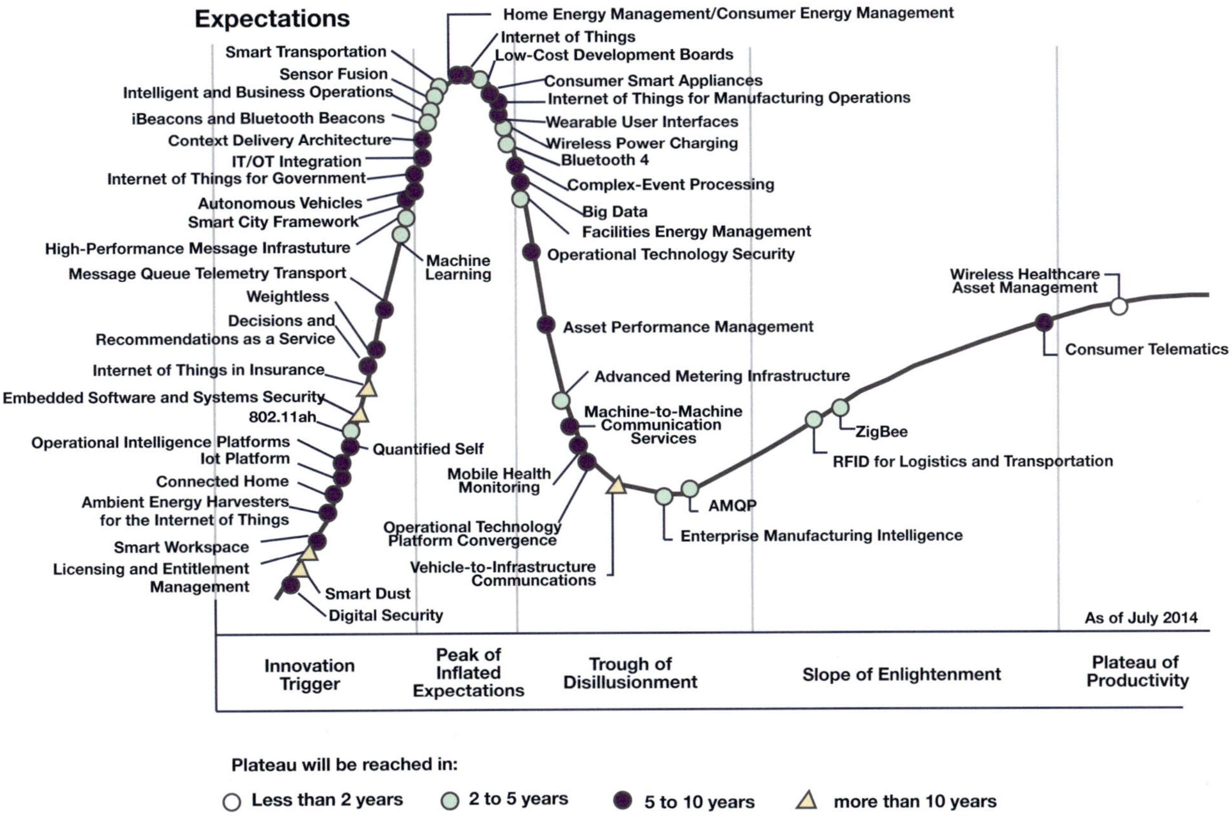

Source: [11]

1.2 WEARABLE DEVICES

The wearable devices most adopted so far are in the fitness, medical and security sectors, while devices for other industry sectors are just emerging. Such interactive products are presently regarded as futuristic or as novelties, but it is expected that such devices will be considered as mainstream fairly soon.

Vandrico's comprehensive database on wearable devices shows how rapidly interest in the technologies increased in 2014 (see Figure 7).

FIGURE 7: GLOBAL INTEREST IN WEARABLE TECHNOLOGY (VANDRICO)

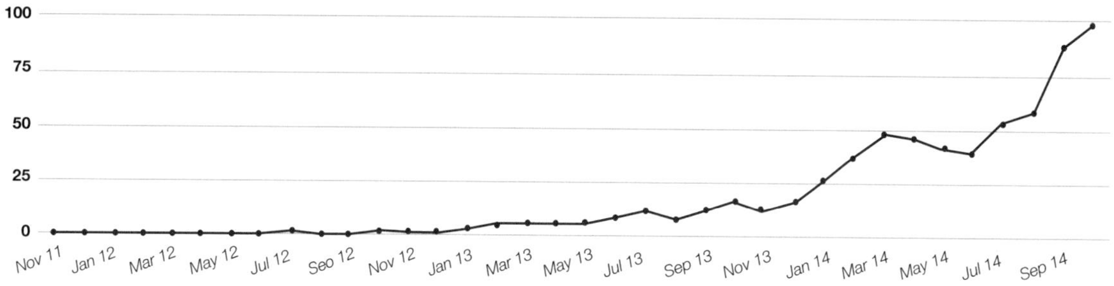

Source: [21]

According to Vandrico, there are 433 devices servicing the lifestyle and fitness category (see Figure 8).

FIGURE 8: WEARABLE DEVICES BY CATEGORY (VANDRICO)

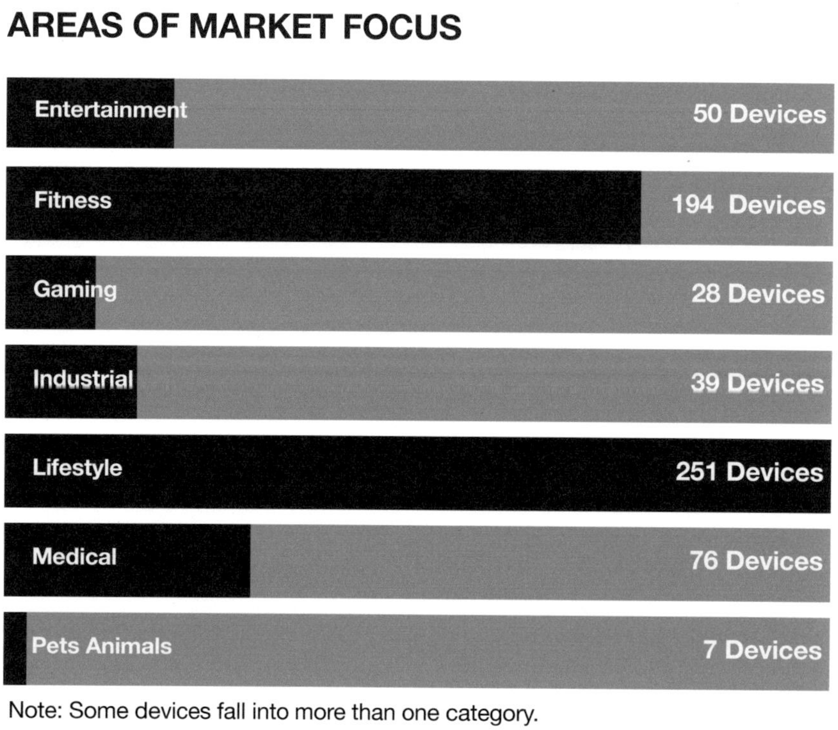

Source: [22]

Figure 9 shows various wearable products, sorted by application sectors [22].

FIGURE 9: WEARABLE DEVICES BY CATEGORY

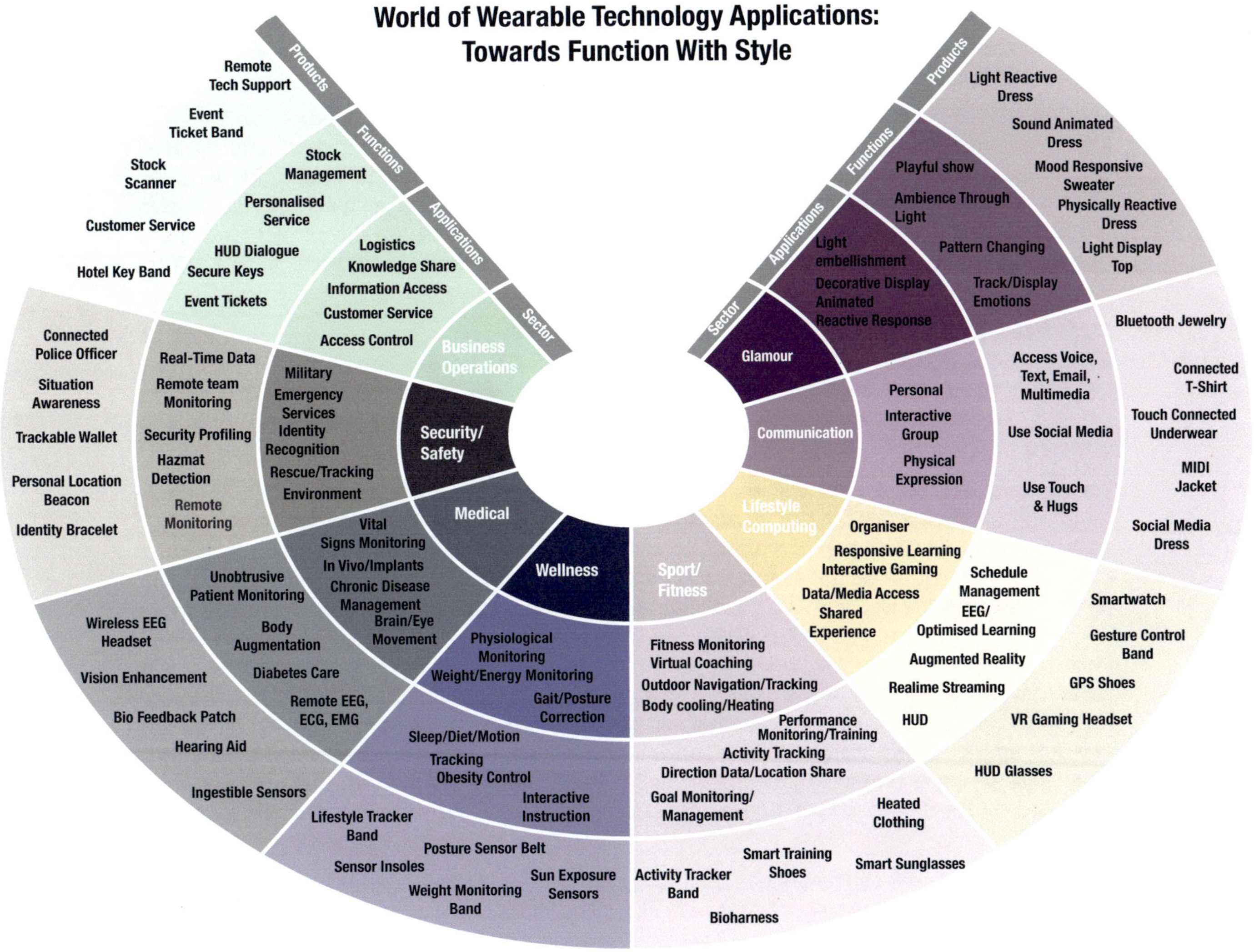

Source: [23]

The recent challenge from Intel 'make it wearable' (MIW) was won by Nixie, a wrist-worn flyable camera designed for the selfie-taking adventurer market. The Intel Project Blog reads:

With traditional cameras, shooters have to interrupt the moment to take pictures, and the controls require manual manipulation. Nixie puts you front and centre in your photos and videos, without requiring your hands or your attention. [24]

Over 1600 experts were asked the following question regarding embedded devices and the Internet/Cloud of Things: "As billions of devices, artefacts, and accessories are networked, will the Internet of Things have widespread and beneficial effects on the everyday lives of the public by 2025?" Some 17% responded with 'no', but the large majority (83%) responded with 'yes'. Elaborating, they said that wearable connected devices will become less visible, and embedded in our bodies, like implants [25].

This thinking is aligned with developments in the healthcare industry. Medical-device giant Medtronic envisions implantable devices that will shift medicine from reactive, symptom-based diagnosis to proactive early detection and prevention. Sensors will monitor vital signs remotely, and the company believes that eventually, people that do not to have such sensors installed in the body will be seen as neglectful [26]. Newdeal Design is working on the project Underskin; developing implanted digital tattoos that will give insight into wellness indicators, and will be supported by identity authentication [27].

Market researcher Gartner also expects 30% of wearable technology to be 'unobtrusive to the naked eye' by 2017 [11]. Google's Project Jacquard (ATAP team) has partnered with jeans manufacturer Levi Strauss to weave conductive threads into a fabric to create interactive textiles [28] [29]. The fabric becomes touch-sensitive (in effect a touch screen) and programmable to gestures.

The question that is not often asked, however, is: "How do we feel about living with technology on (or in) our bodies 24/7?" [30]. A recent study with 200 female Fitbit wearers revealed they considered the device as part of themselves, not as an external technology. The study found that 89% were wearing the device almost constantly, only taking it off to recharge the batteries. The device motivated the wearer to structure their day in order to increase the number of steps when walking, by choosing longer routes (91%) and to increase the amount of weekly exercise (95%). Moreover, 56% increased their walking speed to reach their Fitbit targets faster, and 76% of the wearers were motivated to make healthier food choices (smaller portions and fewer takeaways). However, some women also reported that without their Fitbit they felt 'naked' (45%), others felt that their activities were 'wasted' (43%), and yet others (22%) felt even less motivated to exercise. Also, 79% felt under pressure to reach their daily targets, while 59%) felt their daily routines were controlled by the device (and 30% even considered Fitbit as their 'enemy' that made them feel guilty) [30].

Wearable technologies can have a positive impact on our lives, but we need to be aware that if we invite technology onto or into our bodies, we also need to be willing to share everyday decision-making as to what to do. There are early signs of a technology takeover, leading to questions about the sustainability of the current relationship.

There are examples of wearable technology being used as monitoring indicators, fulfilling an interesting purpose for which they were not originally intended. The 6.0 Richter-scale earthquake that hit the Southern Napa Valley in California in August 2014, is one example.

Due to the quake, 93% of Jawbone Up (fitness monitor) wearers were woken up in cities less than 15 miles from the epicentre, and 45% were kept awake all night. As can be seen in Figure 10, the location of the epicentre of the earthquake could be identified with some precision, based on the location of the affected wearers [31]. In effect, the wearable fitness monitor turned into a perceived earthquake-intensity sensor, and collective data could be used to create a map.

FIGURE 10: PROXIMITY TO EARTHQUAKE EPICENTRE AFFECTED SLEEP OF JAWBONE UP WEARERS

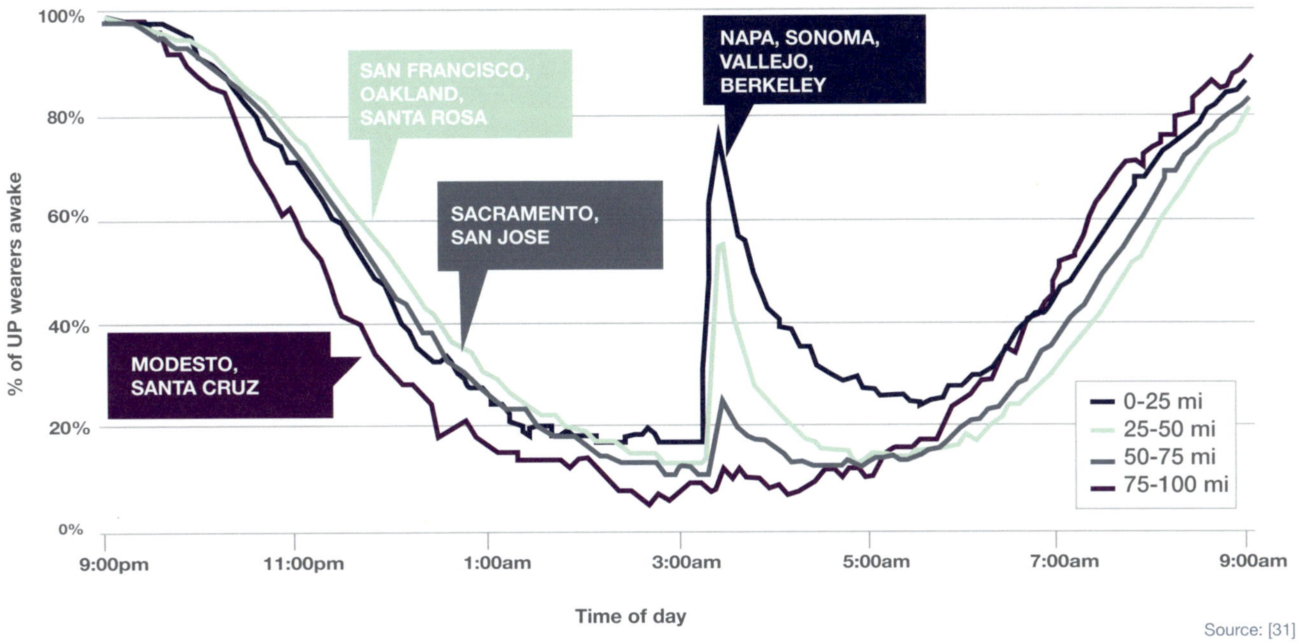

Source: [31]

AUGMENTED REALITY HEADSETS AND EYEWEAR

The global smart glasses market is expected to grow with a CAGR of 194.2% between 2014 and 2019, according to TechNavio's Analysts [32]. More details on selected smart glasses can be found in a list compiled by Wareable (April 2015) [33]. Inbar covers the topic more extensively in his "Smart Glasses Market report 2015: towards 1 billion shipments" [34] (see Figures 11–14).

FIGURE 11: SMART GLASS MARKET – PROGRESS AND MIND SHARE

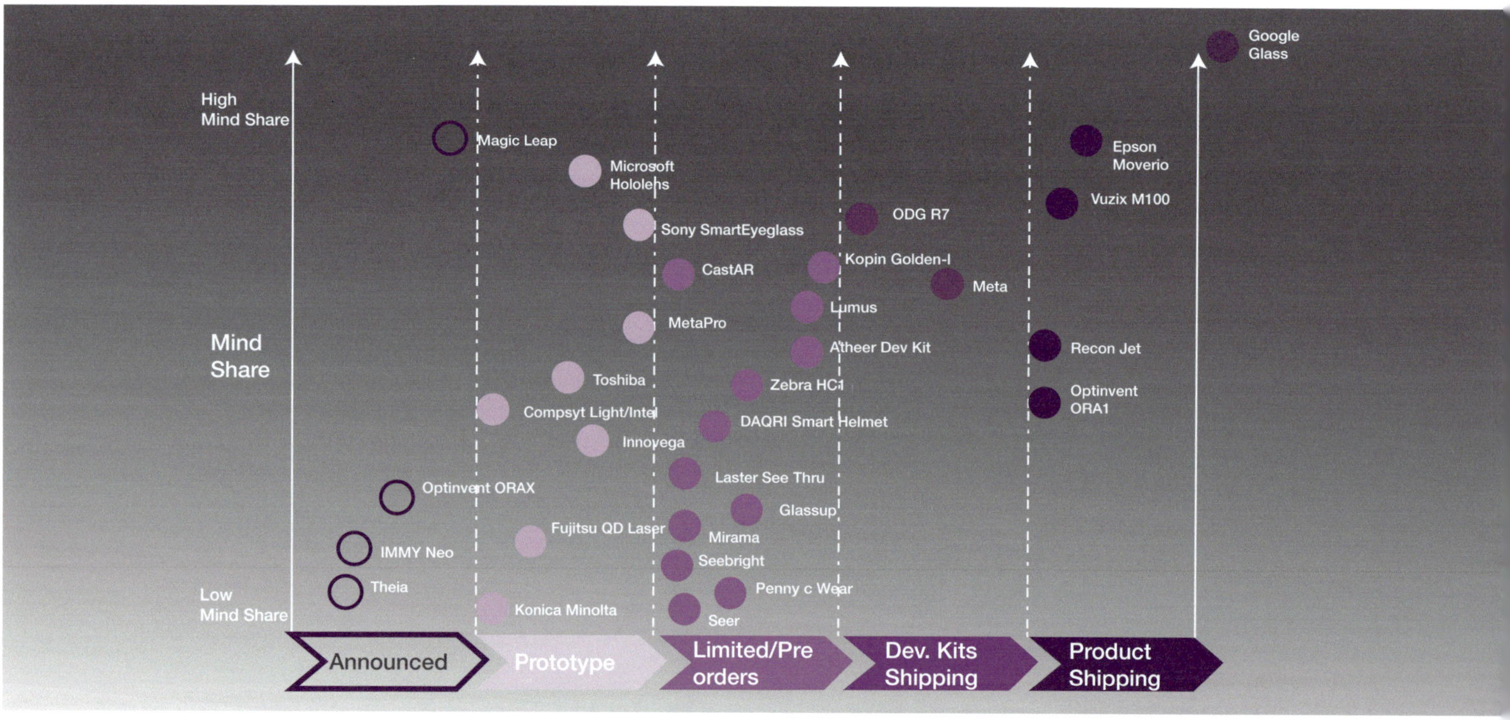

Source: [34]

FIGURE 12: VARIOUS SMART GLASSES

Source: [34]

FIGURE 13: SMART GLASS ADOPTION, 2014–2024

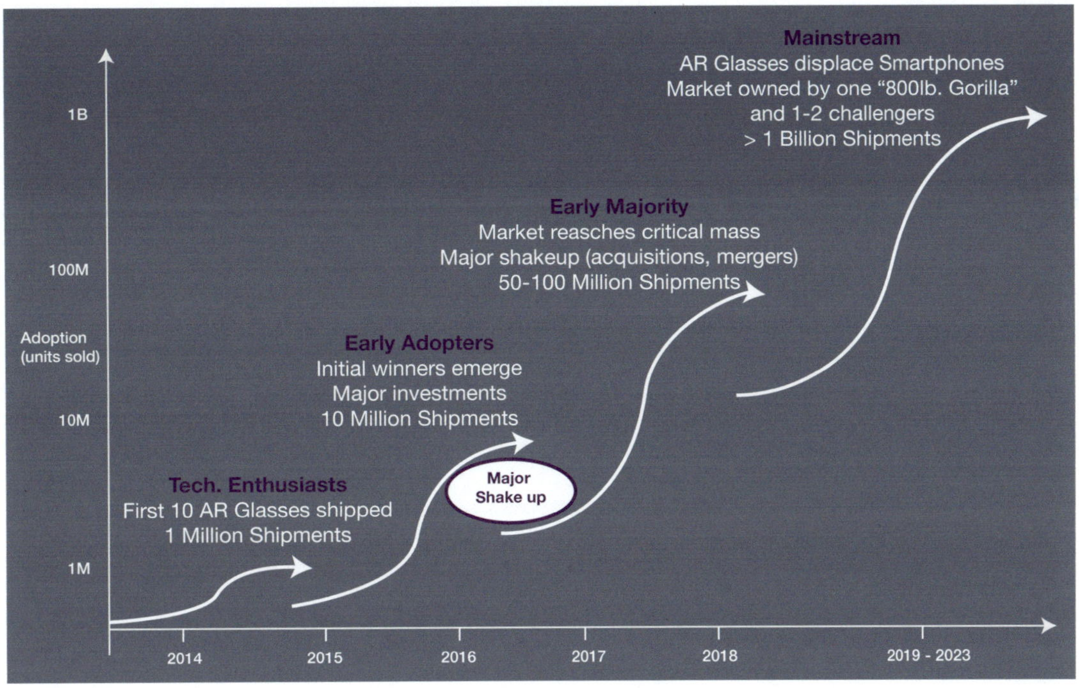

Source: [34]

FIGURE 14: FUTURE SMART GLASS TECHNOLOGIES

Future Generations Breaking the 40 degree Field of View Barrier

Dual Contact Lens-Glasses	Innovega
Direct Retinal Projector	Brother, Fujitsu QDLaser, Eyeway, Composyt Light
Transparent OLED	LusoVu
Pinlight Displays	Nvidia, UNC
Dynamic Lightfield, Vibrating Optic Scanner	Magic Leap
Laser Combiner	Fukui

Source: [34]

Telsyte's Australian digital consumer study of 2015 found fewer consumers than expected were purchasing smart eyewear due to the high prices, lack of local availability, and lack of very useful apps. The smart eyewear apps available, which are related to navigation, and to health and fitness, are both areas that can be addressed with smartphones and other wearable devices, however [35].

Dr Laurie Rothman was among the first in the USA to use Augmedix, a product with software that enables Google Glass for medical record-keeping. In April 2016, Augmedix was reported [36] to have raised US$40 million, focusing on a business application for Google Glass, which had been discontinued for the consumer market. Augmedix saved Dr Rothman hours of data entry and post-exam transcription. Dr Rothman's assertion that "I'm not a techie at all and I love them", indicates the future viability for applications in the medical field [37]. Another medical application for Google Glass is MedRef, built during a Hackathon, which gives access to patient folders via facial recognition and facilitates medical voice notes and photographs [38] [39].

Noteworthy is the Microsoft Hololens, which encompasses holographic computing [40]. Developer kits have been shipped since March 2016 at a cost of US$3000 each to selected developers in the US and Canada [41] [42]. Applications reach beyond gaming: NASA is using Hololens to visualise data from Mars [43], and architects are visualising building plans holographically on the building site [44] (see Figure 15).

In addition, Google is developing smart contact lenses with built-in cameras. The camera is very small and sits near the edge of the contact lens where it does not obstruct the vision; and being part of the contact lens, the camera naturally follows the gaze of the wearer. Applications could include being part of a bionic eye system for the blind, and to enable superhuman powers such as night vision, infrared vision, digital zoom, colliding object alerts etc. [46]. Google also holds a patent where blood sugar levels can be monitored by a 'smart contact lens' [47]. Samsung, moreover, has patented a contact lens with built-in camera that can be operated by blinking [48].

Researchers are even considering future low-cost augmented reality contact lenses for animals; such as applications for cows to see virtual fences [49].

Further, Innovega is developing smart glasses, sponsored by DARPA and a major electronics manufacturer. Innovega's current display has 40 times the screen area of Google Glass, at six times the resolution. The system works by combining contact lenses and glasses (iOptik) [50] and will become commercially available once FDA approval has been granted [51].

FIGURE 15: HOLOLENS AND HOLOGRAPHIC COMPUTING

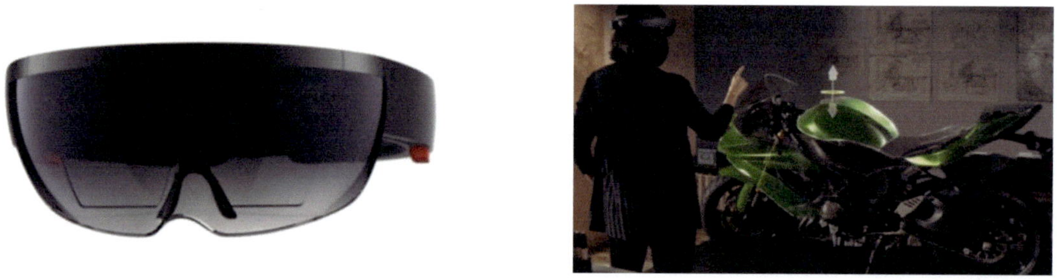

Source: [45]

FIGURE 16: DAQRI SMART HELMET

Source: [52]

Topcon is working with DAQRI to create wearable technology (smart helmets) for surveying, construction, engineering and mining professionals. It aims to make work safer through augmented reality and to give users a hands-free tool that can be used onsite.

An Indiegogo crowdfunding initiative helped launch 'Skully' – a fighter-jet-like motorcycle helmet with heads-up (HU) display for road layout, GPS mapping, speed and distance metrics, and an 180-degree rear camera view [53]. Skully has been called an industry game-changer, greatly improving safety for motorcycle riders. The company started shipping helmets in December 2015 [54] but encountered severe start-up growing pains [55] that need to be overcome for the product to revolutionise the helmet market.

FIGURE 17: SKULLY MOTORCYCLE HELMET WITH HUD

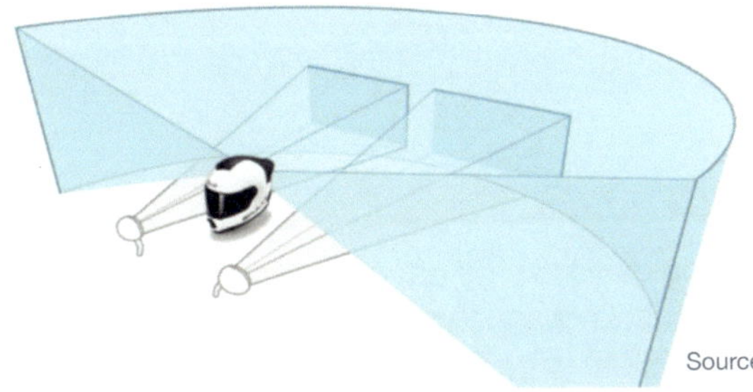

Source: [56]

Smart glasses are currently in a transitional stage and appear to be moving from consumer curiosities to professional tools. The consumer version of Google Glass was discontinued in 2015; partly due to the controversies Google Glass wearers encountered in public. The main objections were privacy concerns due to the built-in camera (a simple slider to cover the camera lens when not in use could have solved many of these problems). Google is now focusing on developing commercial smart glass applications for the healthcare Industry in particular.

VIRTUAL REALITY

Virtual Reality (VR) headsets provide the wearer with fully immersive experiences. The market is forecast to grow at 99% CAGR between 2015 and 2020. It is estimated that more than 26 million VR head-mounted displays will be shipped in 2020. For VR headsets to go mainstream, the devices have to be tethered to gaming consoles such as Sony's project Morpheus PlayStation or the Microsoft Xbox. Sony is set to release its gaming VR headset for PlayStation, priced at US$399, in October 2016; a cost significantly lower than the Oculus Rift version, selling at US$599. By launch date, 50 Playstation games are expected to be ready for use with the VR headset (see Figure 18) [57].

FIGURE 18: GLOBAL GROWTH OF VR HEAD-MOUNTED DISPLAYS

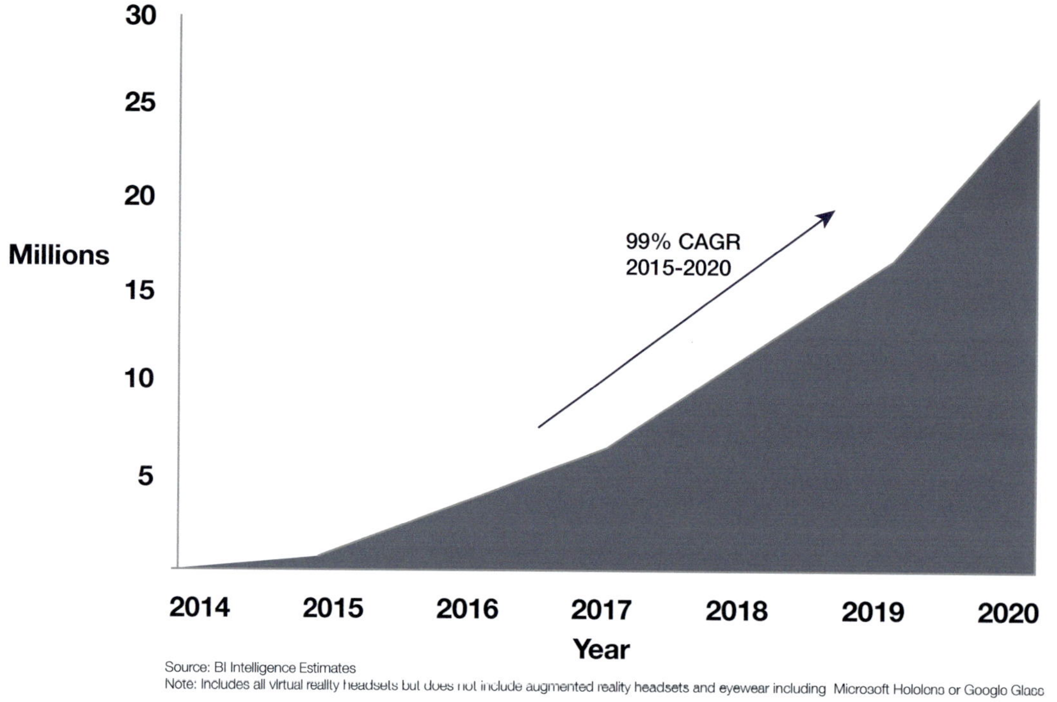

Source: [58]

SMARTPHONES

Wearable devices very often work only when connected to a smartphone. Smartphones act as a central hub, a connected mobile computing system in our pockets, and are now available for as little as US$40 [59].

More and more sensors can now be connected to smartphones, bringing down the cost of the formerly expensive instrumentation. Examples include FLIR ONE, which is an add-on that allows the user to image thermal infrared with the smartphone [60]; EinsteinTM is a lab for students with multiple scientific-sensor extensions that is operated from a smart tablet [61]; and the open-platform Simband is a watch with wristband sensor modules that collects biophysical data [62].

Google's ATAP Project Tango has developed a mobile device that builds up an accurate 3D model of the immediate surroundings by advanced computer vision, image processing and special-vision sensors; use applications will include everything from navigation to AR gaming. Google announced the release of its Tango smartphone in partnership with Lenovo midyear 2016 [63] [64]. The ATAP team is working on Project Ara, an open-module mobile phone, where users choose from a variety of sensors, displays and battery options. Faulty parts can be exchanged instead of discarding the whole mobile device [65].

Smartphones will increasingly be used to facilitate payments. An increase of 1000% of users completing contactless in-store payment transactions at least once a month was estimated for 2015 [66]. According to Forrester Research, US$3.7 billion was transacted by in-person mobile payments in the USA in 2014 [67]. A simple thumbprint on the phone facilitates a speedy transaction (iPhone 6 users only); however, many stores are still lacking the check-out terminal or willingness to accept the payment [68].

The mobile value chain, which directly employs 11 million people, was worth almost US$3.3 billion in 2014., and in the six countries evaluated by Varas [59], the mobile contributed US$1.2 trillion or 24% of GDP (11% in the case of South Korea) in the same year.

According to the International Data Corporation (IDC) report [69], the majority of smartphone users operate on the Android platform. Google Play and the Apple APP store are go-to places for over 96% of smartphone users (see Figure 19).

FIGURE 19: SMARTPHONE OPERATING SYSTEMS

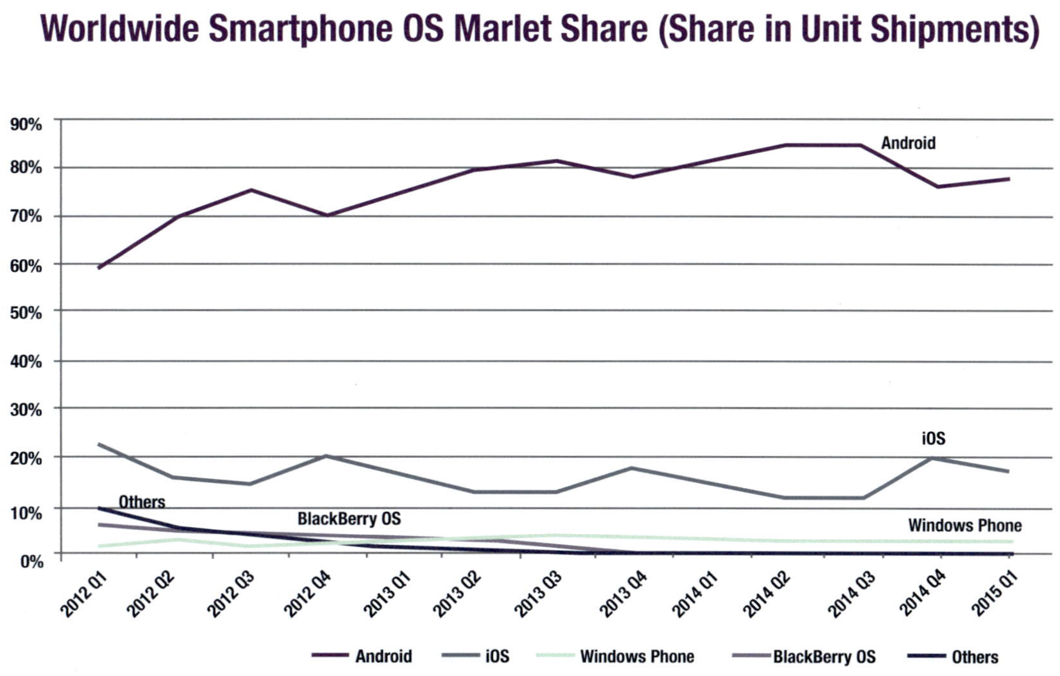

Period	Android	iOS	Windows Phone	BlackBerry OS	Others
Q1 2015	78%	18.3%	2.7%	0.3%	0.7%
Q1 2014	81.2%	15.2%	2.5%	0.5%	0.7%
Q1 2013	75.5%	16.9%	3.2%	2.9%	1.5%
Q1 2012	59.2%	22.9%	2.0%	6.3%	9.5%

Source: IDC, May 2015

1.3 NETWORKS/CONNECTIVITY

Digital services are having enormous economic and social impact. The digital economy is growing by 10% per year. However, there are constraints that need to be addressed, such as the slow adoption of long-term evolution technologies, particularly in Europe, that optimise the speed and function of mobile devices. Areas affected include: spectrum scarcity, disputes over internet protocol interconnection agreements, access, and costs in emerging and rural markets. Furthermore there is a need to address the policies and regulations to accommodate international interoperability [70].

About 3.4 billion people are connected to the internet today (almost half of the world's population) [71]. The average mobile user cost per megabyte decreased by 99% between 2005 and 2013, while mobile data-transmission speeds have skyrocketed: 4G networks are now 12,000 times faster than 2G networks (see Figures 20 and 21) [59].

FIGURE 20: EVOLUTION OF MOBILE PHONE NETWORKS

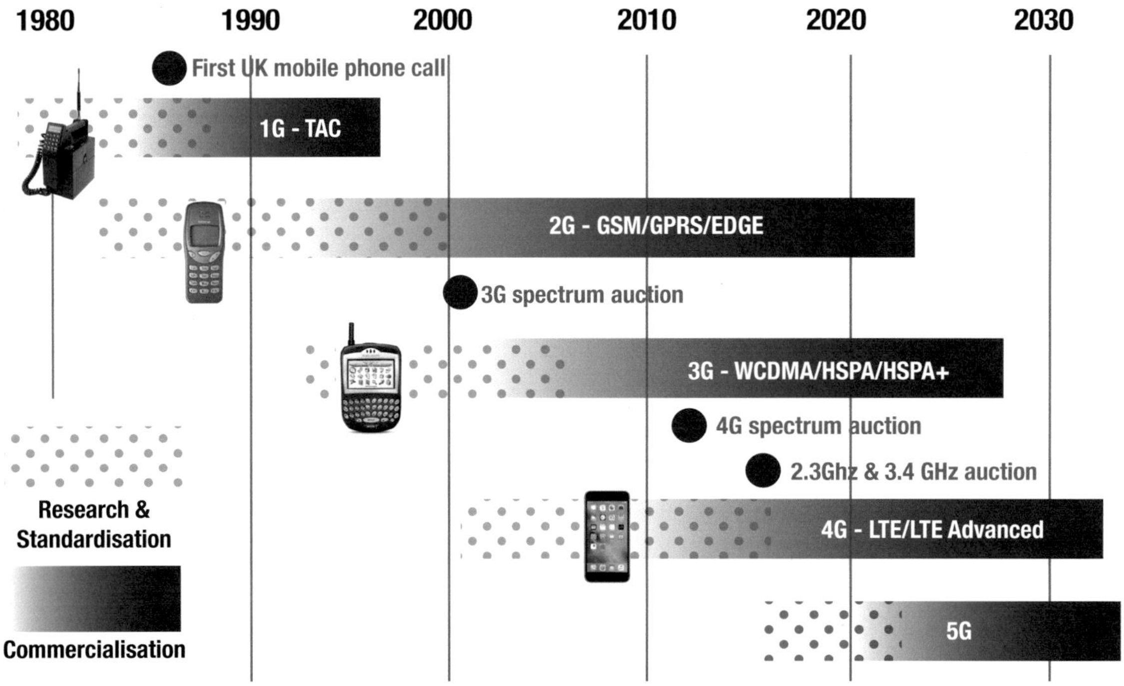

Source: [58]

FIGURE 21: VOICE/DATA AS MEASURED BY ERICSSON

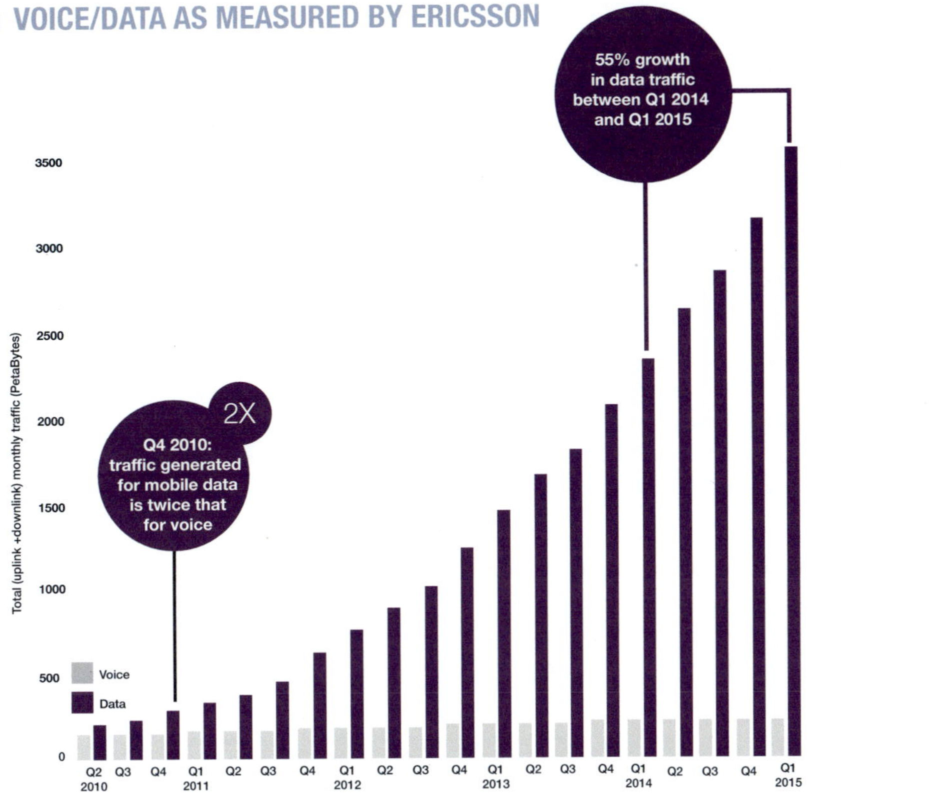

Source: [72]

The Ericsson report also shows how Asia Pacific will consume almost half of the global smartphone traffic by 2020 (see Figure 22).

FIGURE 22: SMARTPHONE DATA TRAFFIC PER REGION

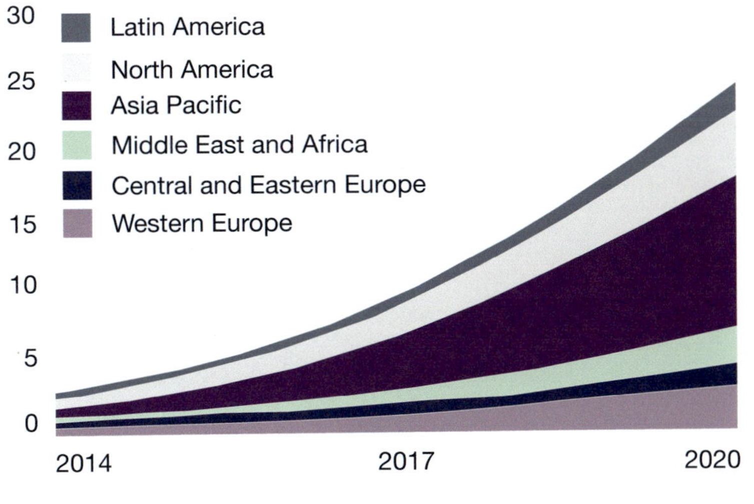

Source: [72]

ASIA PACIFIC WILL GENERATE 45 PERCENT OF TOTAL SMARTPHONE TRAFFIC BY THE END OF 2020

"By 2020, monthly smartphone data consumption per active subscription in North America (14GB) will be 1.5 times that of Western Europe (9.5 GB) and 3.5 times that of Asia Pacific (4GB). However, The Asia Pacific region will have the largest share of total smartphone traffic in 2020, due to growth in the number of subscriptions." [72]

In areas where no conventional networks are available, satellite phones enable communication, although data connections are usually slow. Iridium Go! is a portable hotspot (small, 295g) that turns most smartphones into a satellite phone (speed: up to 2.4kbps) [73].

Existing satellite communications solutions include Inmarsat (BGAN & BGAN M2M), Iridium and Global Star. In October 2015 the first of two Australian NBN communication satellites from Sky Muster (25 Mbps download, and 5 Mbps upload speed) was launched to add capacity to the heavily contested Interim Satellite Service (ISS). In May 2016 [74] NBN satellite Sky Muster became fully operational, and demand for the new satellite is already over capacity.

Where conventional networks are available, the average global network speed (IPv4) is 6.3 Mbps, with South Korea in the first position at 29.0 Mbps (see Figures 23 and 24).

FIGURE 23: GLOBAL TOP 10 – AVERAGE CONNECTION SPEED BY COUNTRY

Global Rank	COUNTRY/REGION	Q1 2016 Avg. Mbps	QoQ Change	YoY Change
	Global	6.3	12%	23%
1	South Korea	29.0	8.6%	24%
2	Norway	21.3	14%	68%
3	Sweden	20.6	8.3%	32%
4	Hong Kong	19.9	19%	19%
5	Switzerland	18.7	12%	25%
6	Latvia	18.3	9.8%	33%
7	Japan	18.2	4.6%	20%
8	Netherlands	17.9	5.5%	20%
9	Czech Republic	17.8	12%	31%
10	Finland	17.7	6.9%	30%

Source: [75]

FIGURE 24: APAC REGION: AVERAGE CONNECTION SPEED

Global Rank	COUNTRY/REGION	Q1 2016 Avg. Mbps	QoQ Change	YoY Change
1	South Korea	29.0	8.6%	24%
4	Hong Kong	19.9	19%	19%
7	Japan	18.2	4.6%	20%
13	Singapore	16.5	19%	29%
21	Taiwan	14.8	15%	46%
39	Thailand	10.8	16%	49%
43	New Zealand	10.5	13%	25%
48	Australia	8.8	7.7%	15%
74	Malaysia	6.4	22%	49%
83	Sri Lanka	5.4	13%	12%
89	Vietnam	5.0	31%	59%
94	Indonesia	4.5	16%	110%
100	China	4.3	3.3%	15%
113	Philippines	3.5	10%	24%
114	India	3.5	24%	55%

Source: [75]

In Australia, the Australian Bureau of Statistics (ABS) found that in Q4, 2015 98% of data downloads were via broadband internet. The volume of data downloaded (1.67 million terabytes) increased by 50.4% in just one year [76].

There are concerns that Australia's performance in broadband speeds has slipped over recent years. Figure 25 shows the ranking as documented in research by the internet companies Akamai and Ookla.

FIGURE 25: AUSTRALIA'S BROADBAND-SPEED RANKING OVER TIME

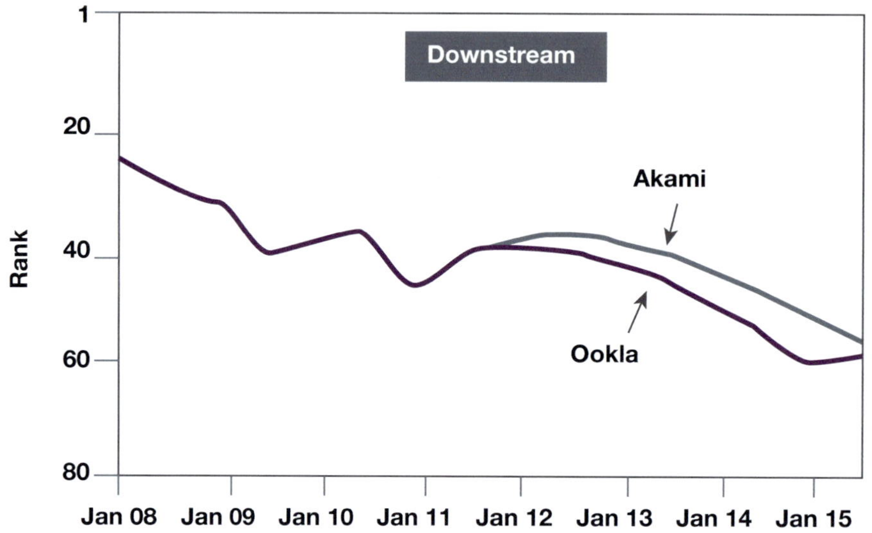

Source: [77]

Australia's current NBN technology approach of 'Fibre to the Node' (FTTN), rather than 'Fibre to the Premise' (FTTP), may be unable to service projected future network speed demands, and consequently may contribute to a continued decline of Australia's position in the world ranking (see Figure 26) [75]

FIGURE 26: PROJECTED NETWORK SPEED DEMANDS VS VARIOUS TECHNOLOGY CAPABILITIES

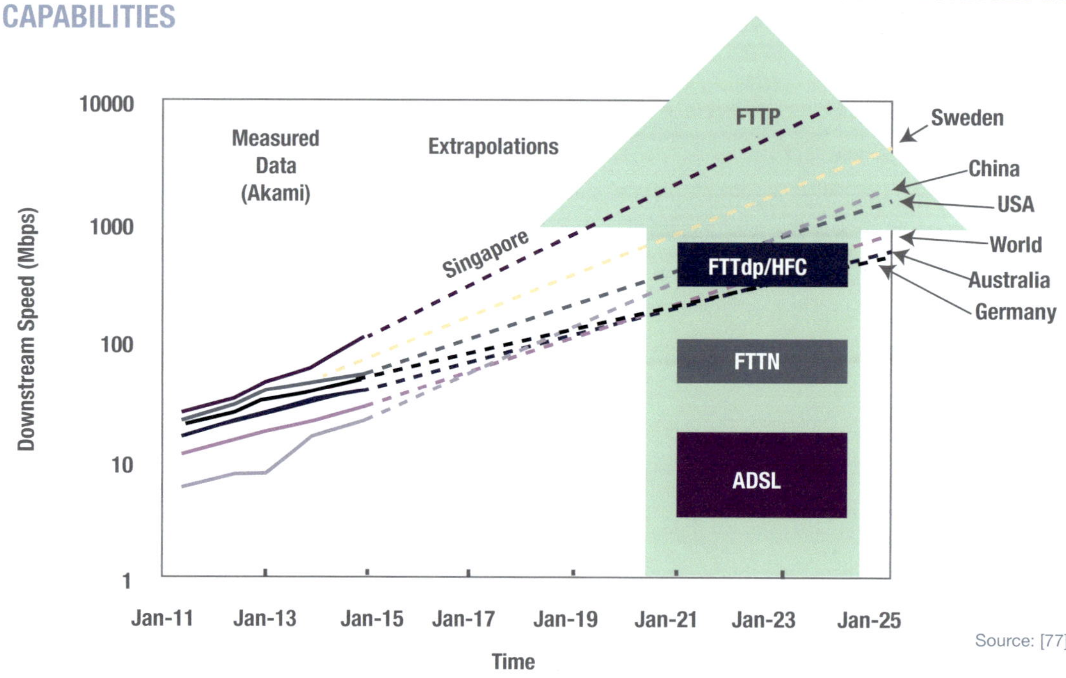

Source: [77]

Broadband network speed and penetration are closely related to National GDP, as documented by Czernich et al. [78] (an increase of 10% in broadband penetration equates to >1% increase in GDP). Further, Kongaut et al. [79] illustrate economic growth scales and broadband speed.

FIGURE 27: MAP OF INTERNET USAGE PER COUNTRY

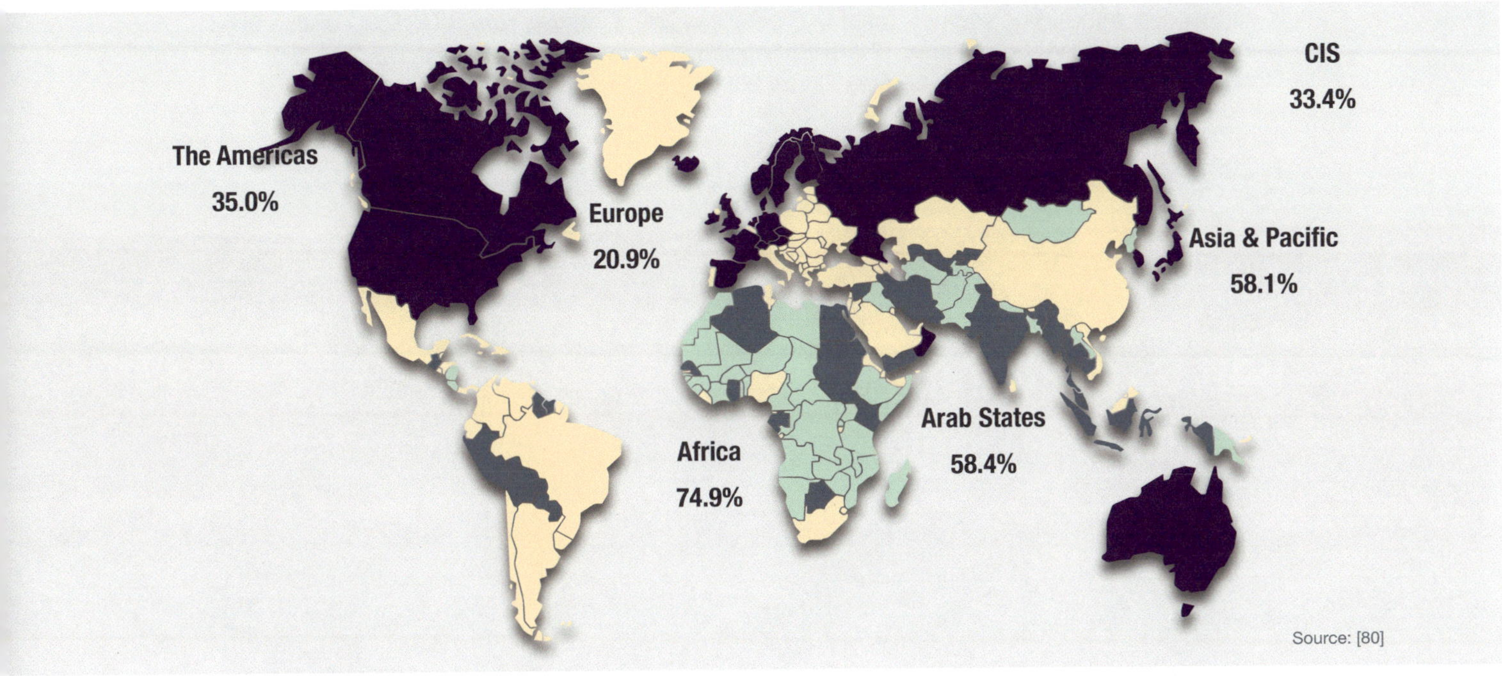

Source: [80]

Percentage of individuals NOT using the internet

- 0-25
- 26-50
- 51-75
- 76-100

By the end of 2016, 3.9 billion people - 53% of the worlds population - is not using the internet.

In the Americas and the CIS regions about one third of the population is offline.

While almost 75% of people in Africa are non-users, only 21% of Europeans are offline.

In Asia and the Pacific and the Arab States, the percentage of the population that is not using the internet is very similar: 58.1% and 58.4% respectivly.

The global telecommunication industry has declared that connecting the remaining two-thirds of the world's population to the mobile internet is a great opportunity, as well as one of the most urgent issues (see Figure 27). The major focus is on the continent of Africa, with its population of 1.1 billion people.

Traditional internet delivery, such as laying underground fibre cables, is often impractical and expensive, often making satellite broadband the preferred option [81]. Companies such as Facebook, Avanti, Google, O3B, Space X, OneWeb and others are currently participating in this second 'space race'; and Mark Zuckerberg is involved in the internet.org initiative. Recently the

Facebook Connectivity Lab reported on the flight-readiness of their Prototype Aquila – a solar-powered high-flying Boeing 737-sized drone capable of staying airborne for up to 90 days – to beam down an internet signal [82]. Alternative approaches being developed include Avanti's satellites that cover large parts of Africa, including South Africa, Zimbabwe, Kenya, Tanzania and Burundi [81].

Sir Richard Branson is supporting the OneWeb initiative (founded by Greg Wyler), to put 648 communications microsatellites in orbit [83]. At the Paris Air Show in 2015, OneWeb announced a partnership with Airbus who will build 900 satellites [84]. OneWeb raised US$500 million from investors including Airbus Group, Bharti Enterprises, Hughes Network Systems, Intelsat, The Coca-Cola Company and Totalplay (see Figure 28) [85].

FIGURE 28: GLOBAL INTERNET SATELLITE CONSTELLATION

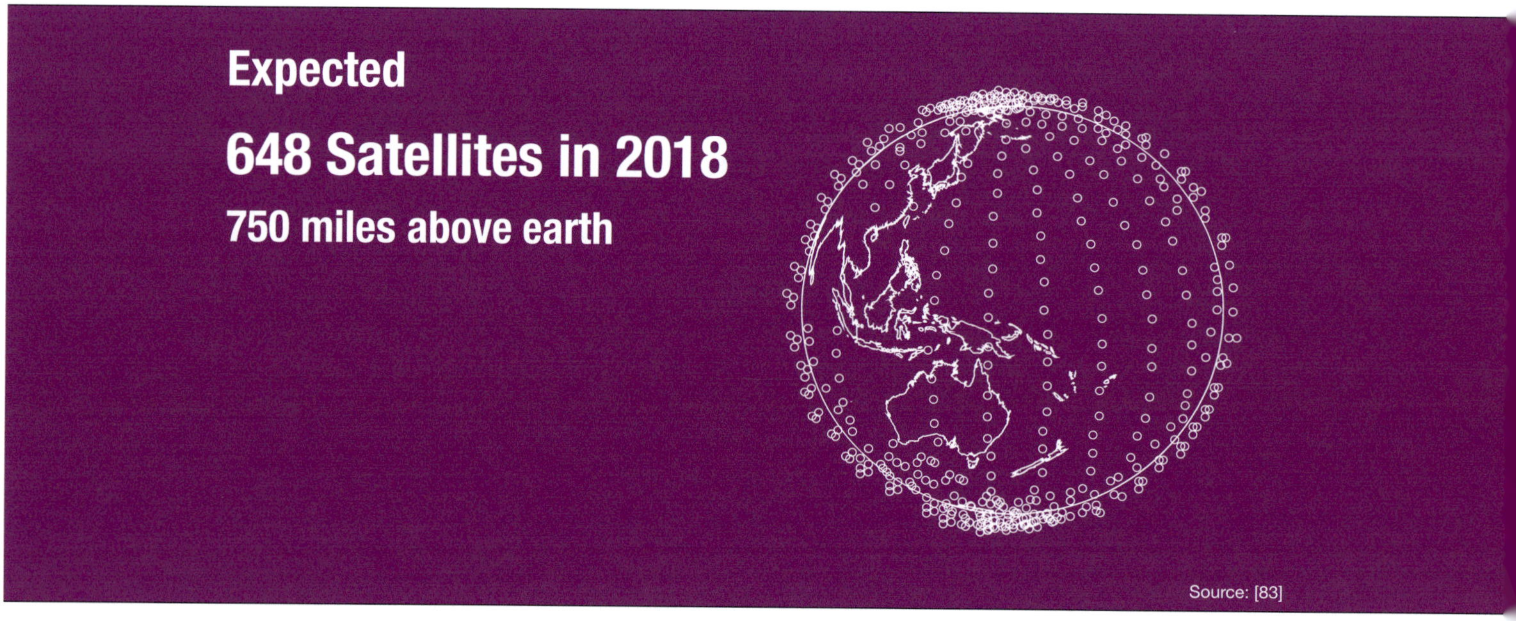

Source: [83]

In a move to deliver the internet by satellite, Google has teamed up with O3B and Elon Musk's SpaceX. Google is also working on alternative solutions such as solar drones carrying small cell towers (Project Titan), and a high-altitude balloon network drifting in the stratosphere (Project Loon). Loon has been tested in multiple countries, and roll- out started in March 2016 in Sri Lanka; thus making it the first country to have universal internet access via the helium balloon system [86]. Furthermore, Google's Project Link is building a fibre backbone in Kampala, Uganda, that is also to be implemented in other African countries [87]. Meanwhile, Samsung has released a report entitled "Mobile internet from the heavens" (Khan, 2015) [88] proposing a fleet of about 4600 satellites in low orbit to ensure global internet coverage with adequate speed and data volume capabilities to meet the demand anticipated by 2028 [89].

1.4 SATELLITES

In 2014, Global Space Activities were worth a total of US$330 billion [90], and the global Space economy grew by 9% [90]. The OECD publication 'Space Economy at a glance 2014' lists the following segments in the Space supply chain:

The space manufacturing supply chain represents conservatively some USD 85 billion globally. This number is probably relatively underestimated since there are institutional programmes in many countries that are the sources of unreported contracts to national space industries (e.g. defence activities). This important segment is often characterised by largely captive markets, since much of the demand for institutional satellites, launchers and ground segment is often directed at national industries. More actors than ever before are involved in supplying space products. As a second segment, services from satellite operators – which own and operate satellites – are included for some USD 21.6 billion (i.e. revenues from the satellite telecommunications operators: fixed and mobile satellite services, satellite radio services, and commercial remote sensing operators). These are important actors, as they have to service governmental and commercial customers outside the space sector (e.g. providing bandwidth, imagery), so they tend to push space manufacturing suppliers for more innovation to respond to market needs at lower cost (e.g. development of broadband via satellite).

Finally, the consumer services include actors, usually outside the space community, which rely on some satellite capacity for part of their revenues. These downstream activities are an integral part of the space economy, although their share is the most difficult to assess, as valuable satellite signals or data need to be tracked in equipment and services. They include direct-to-home satellite television services providers, satnav consumer equipment and value-added services, and very-small apertures terminals providers (e.g. data handling, banking), with revenues estimated at some USD 149.6 billion (see Figures 29 and 30).

FIGURE 29: MAIN SEGMENTS OF THE SPACE ECONOMY

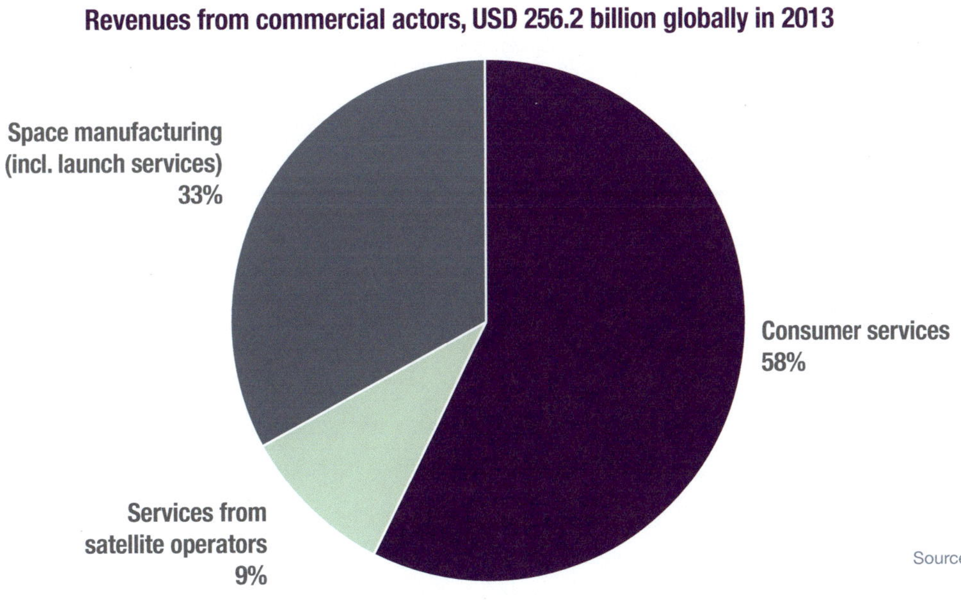

Source: [91]

FIGURE 30: OECD-CALCULATED SPACE BUDGETS OF SELECTED NATIONS

Country	Space Budget in USD millions (PPP), 2013	Budget Per Capita
USA	39332.2	123.2
CHN	10774.6	7.9
RUS	8691.6	61.0
IND	4267.7	3.3
JPN	3421.8	26.9
FRA	2430.8	38.0
DEU	1626.6	20.1
ITA	1223.3	20.7
KOR	411.5	8.2
CAN	395.9	11.5
GBR	338.9	5.3
ESP	302.9	6.7
BRA	259.2	1.3
BEL	244.8	21.9
IDN	142.0	0.6
CHE	133.0	16.6
SWE	122.0	12.7
NDL	110.5	6.5
TUR	104.3	1.4
NOR	89.6	18.5
ISR	89.3	11.1
POL	80.7	2.1
ZAF	76.4	1.5
AUT	73.0	8.6
FIN	53.9	9.9
DNK	38.2	6.9
PRT	32.2	3.0
GRC	30.3	2.7
CZE	25.4	2.5
IRL	25.3	5.6
AUS	24.9	1.1
LUX	17..0	34.5
HUN	8.9	0.9
MEX	8.5	0.1
EST	5.4	4.0
SVK	4.8	0.9
SVN	2.9	1.4

Source: [91]

Source: OECD Calculations based on national data and OECD MEI data.

In Australia, EOS alone contributed direct economic benefits of AUD$496 million in 2015, and AUD$1,694 million are expected by 2025. "The Value of Earth Observations from Space to Australia, 2015" (ACIL Allen Consulting) [92] found that EOS returns approximately AUD$5.3 billion to the Australian economy every year, and the use of EOS services generated around 9000 new jobs in 2015. The report predicts growth of up to AUD$8 billion, with more than 15,000 to be employed as a direct result of EOS services by 2025 (see Figure 31) [92].

FIGURE 31: EOS ECONOMIC BENEFIT TO AUSTRALIA

Activity	Sectors	Impact	Nature of benefit	2015	2025	Potential employment impacts 2015	Potential employment impacts 2025
				$million	$million		
Weather forcasting	Agriculture, Aviation	Economic	Increase in value added	75.8	199.1	903	1474
Ocean observation	Great Barrier Reef	Social/ environmental	Sustainable ecosystem	825.0	1275.0	7593	10146
	Petroleum and Shipping	Economic	Increase in value added	13.3	29.7	11	25
Landscape and land cover monitoring	Government	Social/ environmental	Sustainable land cover	20.7	39.2		
	Property	Economic	Increase in value added	64.4	128.8	384	768
Agriculture	Crops and pastoral, biosecurity	Economic	Increase in value added	84.7	754.7	402	3583
Water management	Water sector	Economic	Increase in value added	39.5	80.0		
	Water resources	Social/ environmental	Enhancing habitat and catchment ecology	14.9	14.9		
Natural disasters	Insured costs plus social costs	Economic	Increase in value added	213.0	495.0		
Mining	Licence to operate	Economic	Increase in value added	5.0	6.7		
Total Economic				495.7	1696.0	9293	15997
Total social/ environmental				860.6	1329.1		

Note: This project scope did not include economy-wide modelling of the direct economic effects on GDP and incomes. The totals are therefore direct effecys and not outcomes for the economy as a whole. Source: ACIL ALLEN

Source: [92]

According to data from the Union of Concerned Scientists (UCS), there are currently 1,419 active satellites in orbit as of 30 June 2016 (see Figures 32–34) [93].

FIGURE 32: SATELLITE QUICK FACTS

Total number of operating satelites: 1,419			
United States: 576	Russia: 140	China: 181	Other: 522
LEO: 780	MEO: 96	Elliptical: 37	GEO: 506
Total number of U.S satellites: 576			
Civil: 12	Commercial: 286	Government: 132	Military: 146

Source: [93]

FIGURE 33: NUMBER OF SPACECRAFT LAUNCHED, 2005–2014

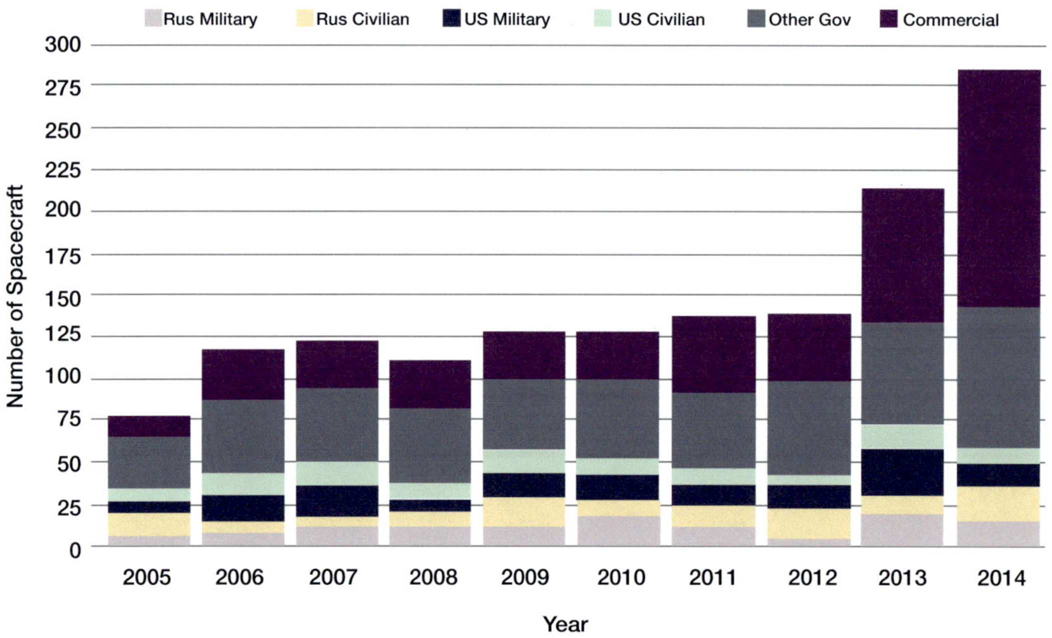

Source: [94]

FIGURE 34: ACTIVE SATELLITE PAYLOADS PER YEAR, 1958–2014

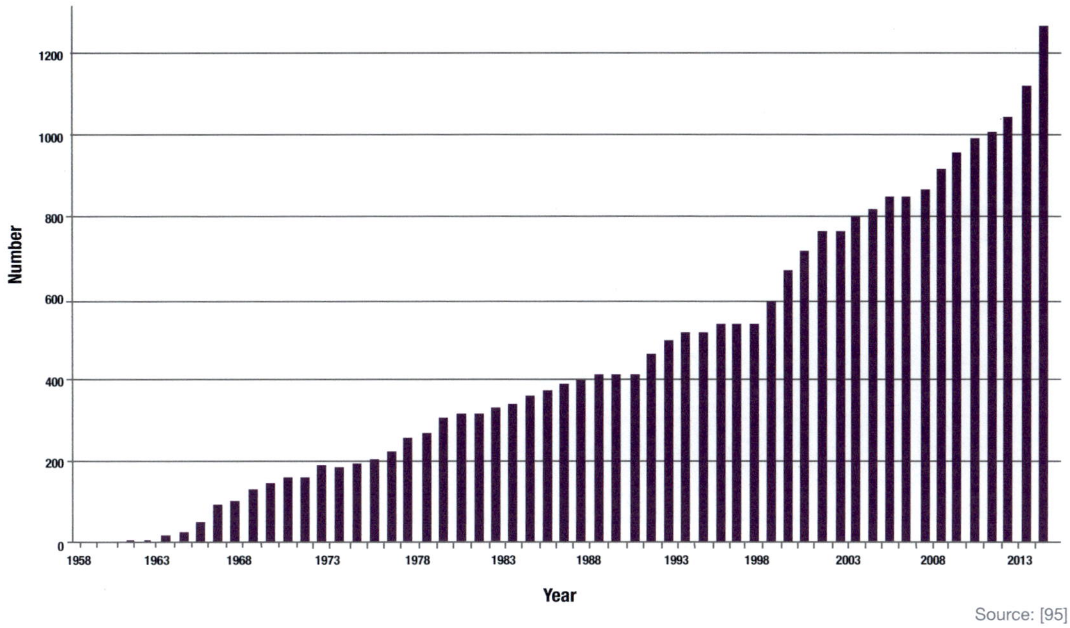

Source: [95]

Figure 35 shows satellite operators by country. The data are from the database maintained by the UCS [93]; 'Other' includes multinational satellites.

FIGURE 35: SATELLITES BY COUNTRY

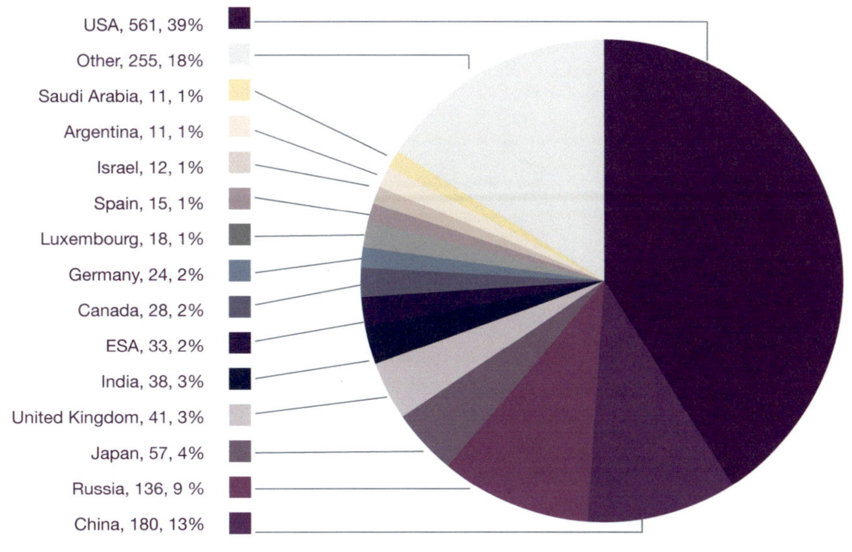

Australia, included in 'Other' in Figure 35, operates seven communication satellites and one shared communication satellite with the US (see Figure 36). Optus is owned by Singtel, a Singaporean company.

FIGURE 36: AUSTRALIAN SATELLITES, ACCURATE AS OF 30 JUNE 2016

NORAD	COSPAR	Satellite Name	Launched	Contractor
20410	1990-002B	Leasat 5 (Syncom IV-5, Leased Satellite F5)	9/01/1990	Hughes
23227	1994-055A	Optus B3	27/08/1994	Hughes
27831	2003-028B	Optus C1	11/06/2003	Space Systems/Loral
29495	2006-043B	Optus D1	13/10/2006	Orbital Sciences Corp.
32252	2007-044A	Optus D2	5/10/2007	Orbital Sciences Corp.
35756	2009-044B	Optus D3	21/08/2009	Orbital Sciences Corp.
40146	2014-054A	Optus 10	11/09/2014	Space Systems/Loral
40940	2015-054A	NBN-1A (Sky Muster)	30/09/2015	Space Systems/Loral

Source: [96]

FIGURE 37: CHANGES IN IMAGING SATELLITE RESOLUTION OVER TIME

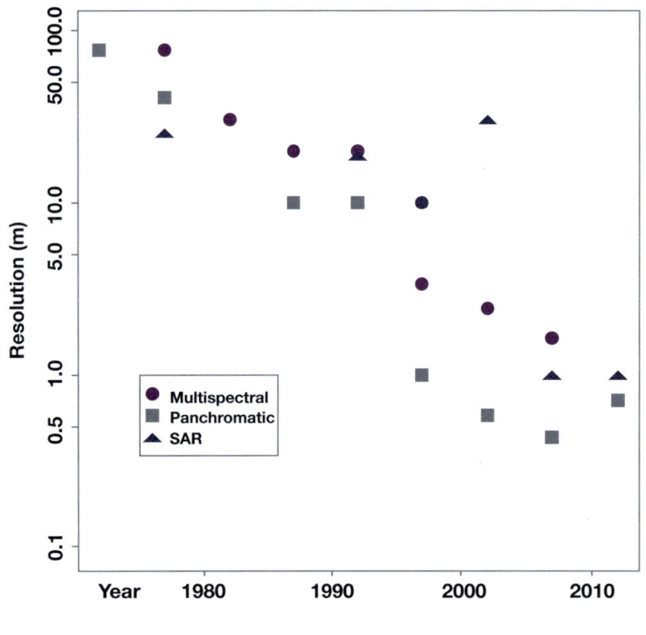

The highest resolution (meters) achieved from any panchromatic, multispectral and/or SAR sensor onboard a near-polar orbiting, land imaging civilian satellite in each 5 year period's launches (the figure does not take individual missions' lifespan into account).

Source: [98]

Until recently, the best commercially available pixel resolution from satellite imagery was 50cm, but the US Government relaxed that restriction on 21 February 2015, and 30cm imagery is also now available [97].

Figure 38 shows the improvement in resolution over time for civil satellites compared to military satellites.[2]

[2] Note: KH=Keyhole: designates several DoD reconnaissance satellite series, such as CORONA, ARGON, LANYARD, and LACROSSE as well as the principal camera system of the Spacecraft

FIGURE 38: MILITARY VS CIVIL SATELLITE RESOLUTION OVER TIME

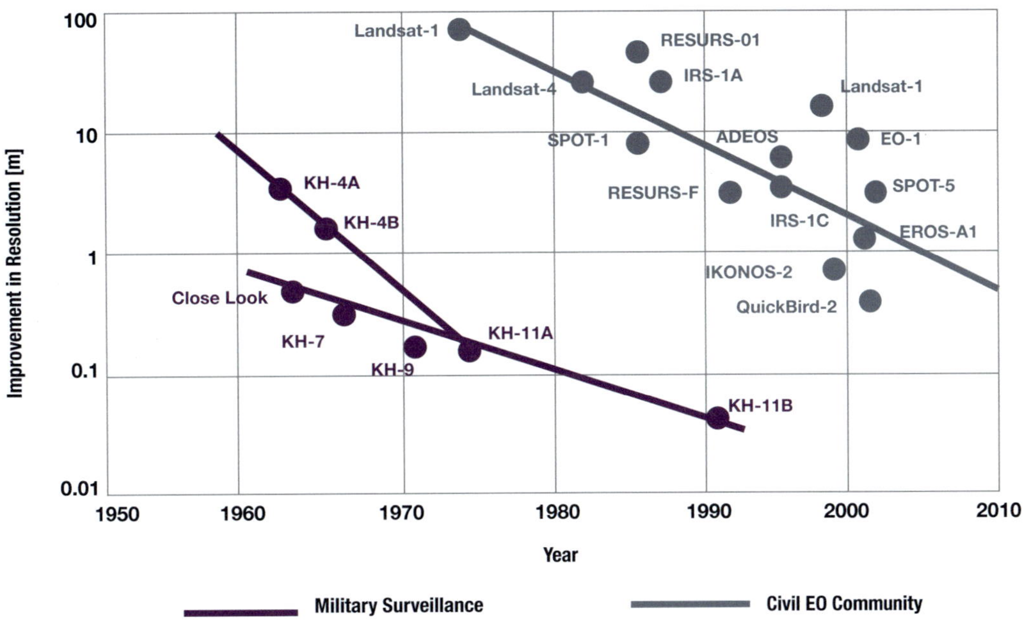

Source: Kramer, 2002 [99].

The fleet of satellites operated by private companies is rapidly expanding as costs to build and launch come down (see Figure 39 for civilian EO satellite launches). In November 2014, Orbital Science launched (from Virginia, USA) 29 satellites at once (a record at the time) in a LEO orbit followed a few days later by the Russian joint-venture Kosmostras that deployed 32 satellites into a similar orbit. The performance of these Cubsat satellites, which measure about 10cm, weigh 1.3kg and use smartphone components, compares well to the traditional technologies for a fraction of the cost [100]. STRaND-1 was the first 'phonesat' to go into orbit, built by Surrey Satellite Technology (part of the Airbus group) out of a Google Nexus One smartphone and a 3-unit CubeSat [100].

QB50 is a network that will launch 50 CubeSats together in a string-of-pearl configuration in 2016. Participant, the University of New South Wales (UNSW) will test the Namuru Space GPS, and demonstrate the use of a rapid-prototyped satellite structure in the space environment amongst other objectives for their CubeSats [101]. Planetlabs builds its nanosats in 'clean-enough' rooms in San Francisco. The company has 75 satellites currently in operation, and plans on launching another 150 nanosats into orbit within the next year or so. Several rocket mishaps during launch have slowed their satellite deployment, however. Planetlabs has also used the Nanosat Cubsat Deployer on the ISS. In July 2015 Planetlabs announced the acquisition of Black Bridge and its Rapid Eye satellite constellation [24]. URthecast broadcasts live imagery from the ISS (International Space Station) [102], and aims to be the largest commercial user of the ISS [103]. Airbus Defence, with its Pleiades and SPOT satellites, has formed a partnership with Hexagon Geospatial to provide access to high-resolution satellite data in smart applications as well as planning future access to TerraSAR-X radar data and digital elevation models [104].

FIGURE 39: MAP OF CIVILIAN IMAGING SATELLITES LAUNCHED, BY COUNTRY

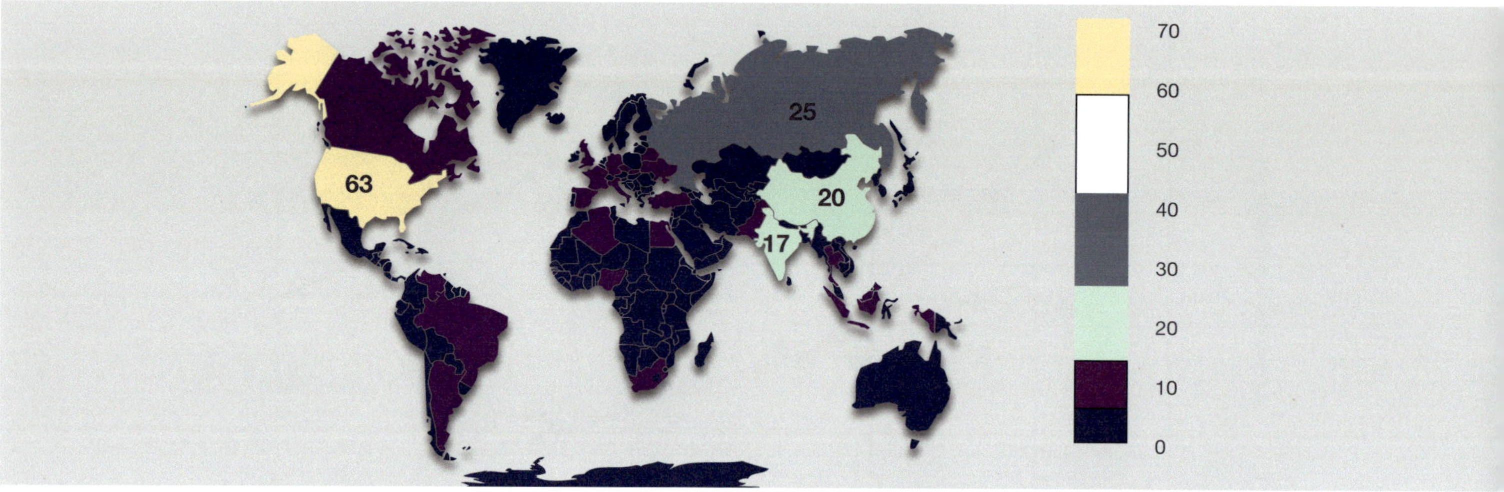

Map showing the total number of near-polar orbiting, land imaging civilian satellites launched by (or on behalf of) different geographical regions between 23rd July 1972 and 31st December 2013. The legend to the right of the map shows the number according to 7 groups; 0; 1-5; 6-10; 11-20; 21-40; 41-60; 60-70. Note that no region falls into the 41-60 category. The numbers of launches made by the top four individual countries (India, China, Russia and USA) are specifically cited - note collectively Europe has launched 30 missions.

Source: [98]

A comprehensive list of EO satellites is referenced in APPENDIX A.

Details on successful launches are recorded in the launch log of 'Spaceflight NOW' [105], and future scheduled launches are noted in the launch schedule [106] of the same website. Richard Langley, GPS World innovation editor and professor in the Department of Geodesy and Geomatics Engineering at the University of New Brunswick (UNB) maintains a database specifically for upcoming GNSS-satellite launches:

"Upcoming GNSS Satellite Launches: GPS World", June 2016 (see Figure 40).

FIGURE 40: UPCOMING GNSS LAUNCHES

Upcoming GNSS Satellite Launches				
System	Satellite	Launch Site	Launch Date (UTC)	Launch Time (UTC)
BeiDou	BeiDou GEO	Xichang	12 June 2016	~15:35
WAAS	Eutelsat 117W B	Cape Canaveral	14 June 2016	14:32-15:17
Galileo	Galileo 15-18	Kourou	17 November 2016	TBD
QZSS	QZS - 2	Tanegashima	NET January 2017	TBD
GPS	GPS 3-1	Cape Canaveral	3 May 2017	TBD

Updated: 10 June 2016

Source: [107]

1.5 LOCATION

The global real-time location system market is expected to grow at a CAGR of 37.41% between 2015–2019 [108]. In the USA alone, GPS contributed US$68 billion to the economy in 2013 (a conservative estimate, as this did not include GPS-timing applications or other similar apps). More than US$26 billion was related to vehicle-location services, US$13.7 billion to precision farming, $11.9 billion to fleet-vehicle telematics, US$11.6 billion to surveying, and US$5 billion to GPS-based guidance of earth-moving equipment [109].

Australia is currently considering updating its datum in a proposed two-stage approach. Stage 1 in 2017 will implement the Geocentric Datum GDA2020 (approximately a 1.8m shift to a new static datum), and Stage 2 will involve the transitioning to the new Australian Terrestrial Reference Frame (ATRF) and a dynamic datum sometime after 2020 [110]. In this way Australia will be well-prepared for an economy and a society that will be increasingly dependent on precision positioning; especially for areas involving mobility such as controlled traffic farming, robotically enabled mines and ports, and autonomous transport.

Global and Regional Navigation Systems (including GPS) are operated by a number of nations (see Figure 41).

FIGURE 41: OVERVIEW OF GNSS SYSTEMS

System	Owner	Coding	Number of satellites	Frequency
GPS	United States	CDMA	32 (at least 24 by design)	1.57542 GHz (L1 signal); 1.2276 GHz (L2 signal)
GLONASS	Russian Federation	FDMA/CDMA	28 (at least 24 by design)	Around 1.602 GHz (SP); Around 1.246 GHz (SP)
BeiDou	China	CDMA	35 (5 GEO, 30 MEO)	1.561098 GHz (B1); 1.20714 GHz (B2); 1.26852 GHz (B3)
Galileo	European Union	CDMA	30	1.164–1.215 GHz (E5a and E5b); 1.260–1.300 GHz (E6); 1.559–1.592 GHz (E2-L1-E11)
NAVIC	India	CDMA	7	1,176.45 MHz (L5); 2,492.028 MHz (SPS)
QZSS	Japan	CDMA	4 (7 in future)	L1 (1575.42 MHz); L2 (1227.6 MHz); L5 (1176.45 MHz); LEX (1278.75 MHz) compatible with E6;

Sources: UNOOSA; Wikipedia; Glonass-IAC; UCSUSA.
Note: Number of satellites: active and planned satellites in constellation

A comprehensive list of GNSS satellites is referenced in APPENDIX B.

A team of researchers from the CRC for Spatial Information (CRCSI), Curtin University of Technology, RMIT University, the University of New South Wales and Geoscience Australia have made a number of helpful advances in the use of GNSS capabilities in recent years. These include the successful testing of JAXA QZSS signals on a robotically controlled tractor; the trial of which achieved a positioning accuracy to around 5cm. It also enabled the use of the tractor's power take-off unit for ploughing and other functions, using the communications channel that is also a feature of QZSS [111].

The CRCSI's research program addresses issues of multi-constellation; the multi-frequency signal integration in support of a new positioning paradigm known as PPP-RTK. Specific challenges include the need to[3]:

- Compute precise, real-time satellite orbit and clock products
- Produce a national 3D ionospheric model with high spatial and temporal resolution
- Integrate signals from all satellite systems
- Eliminate or model the various inter- and intra-system biases
- Perform multi-frequency ambiguity resolution in real-time with high reliability

The program work is creating the algorithms needed to rigorously incorporate a full multi-GNSS processing capability into the ACS. This work has recently reported the following developments[4]:

- Single-point positioning capabilities of the Indian Regional Navigation Satellite System (IRNSS/ operational name: NAVIC) using real-data from a premature configuration and simulated data from its fully operational configuration.
- Analysis of carrier-phase IRNSS data together with a study of its potential single-frequency positioning capabilities when combined with GPS (USA), QZSS (Japan) and Galileo (Europe).
- Analysis of GLONASS (Russia) L3 signals that demonstrated ambiguity resolution and positioning performance using only the two new GLONASS satellites currently transmitting the L3 signal.
- The dynamic characteristics of multi-GNSS receiver code and phase biases using mixed receivers from year-long time series from a range of sources.

All these developments are contributing to the creation of Australia's National Positioning Infrastructure that over the next five years, intends to provide a single network solution to any user in real-time anywhere outdoors to around 2cm positional accuracy.

National Space-based positioning, navigation and timing (PNT) are deemed a critical infrastructure by the US Department of Homeland Security. However, the GPS constellation itself is not [112]. DARPA, with its Spatial, Temporal and Orientation Information in the Contested Environments program (STOIC), is working on a brand new global positioning, navigation and timing system that is resistant to jamming and does not depend on satellites. Using miniaturisation, quantum physics, and pulsed lasers, they are looking at novel inertia measurement devices with cold-atom interferometry and affordable chip-scale self-calibrating devices (gyroscopes/accelerometers/ clocks) [113].

In the wake of signal jamming from North Korea, South Korea signed a contract in October 2015 to implement an enhanced national Loran system (eLoran), ensuring uninterrupted positioning navigation and timing services [114]. Integrated eLoran/ GPS-system development is also under discussion in the US$121 million, 5-year e-Navigation program (see Figure 42).

---

[3] Personal communication Phil Collier, Research Director CRCSI; 15 June 2016.
[4] Ibid.

FIGURE 42: SOUTH KOREAN ELORAN SYSTEM

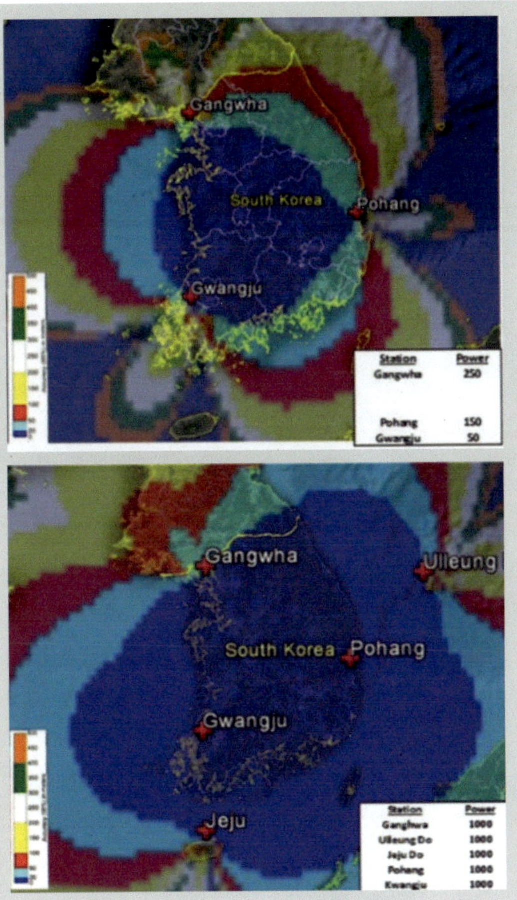

Source: [115]

An Australian innovation by Locata has the potential to replace or augment GNSS signals and overcome the challenge posed by a signal loss in indoor and urban environments, deep open-cut mines, and intended/unintended RFI (Radio Frequency Interference). The Locata-synchronised transceivers form a network and enable centimetre accurate single-point positioning using phase measurements without utilising the differential technique [116] [117].

Locata and Leica Geosystems partnered to develop LocataLite-enabled GNSS receivers for use in Leica's mine-machine automation and mine-fleet management market.

It is likely that Locata technology will become more affordable over time, and may eventually permit farm cooperatives, for example, to establish Locata networks within a region. Miniaturisation over the next few years will bring both the size and price down. It is expected that in five years' time the receivers will be chip size and cost about AUD$10 each[5]. Tested at US White Sands Missile Range, the company's V-Ray antenna delivered location information at 6cm horizontal, and 15cm vertical accuracy to an aircraft flying at 25,000 feet altitude, at a speeds of 550 mph [118]. Consequently, Locata has been selected as the core positioning technology for RPAS research at the NASA Langley Research Center in Hampton, Virginia [119], and in August 2016 the system has been declared operational [120].

[5] Personal communication from Nunzio Gambale, CEO Locata, January 2016.

Figure 43 from R. Mautz [120] shows the accuracies of indoor location methods, and Figure 44 shows the EM frequency ranges used by the various positioning systems.

FIGURE 43: SPATIAL ACCURACIES WITH ESTABLISHED INDOOR LOCATION METHODS

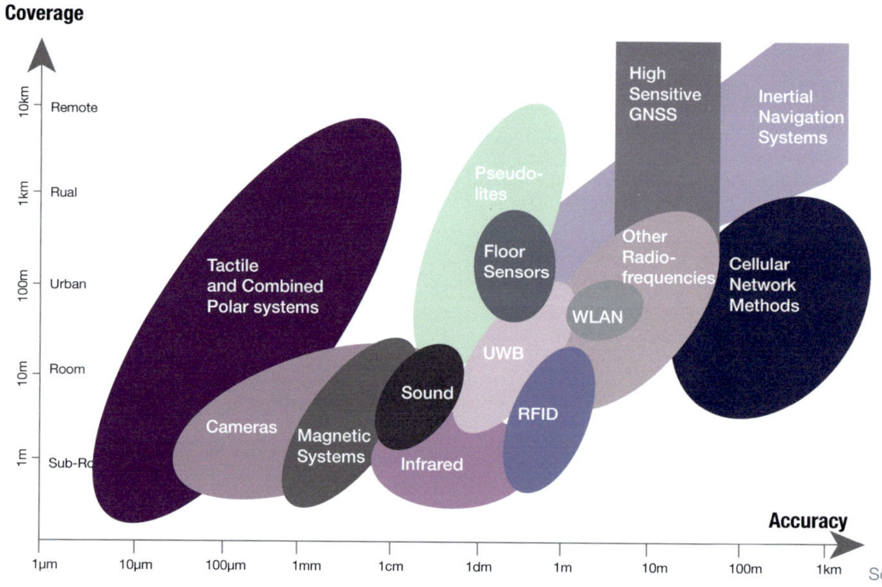

Source: [121]

FIGURE 44: CARRIER WAVELENGTH AND ACCURACY OF INDOOR POSITIONING SYSTEMS

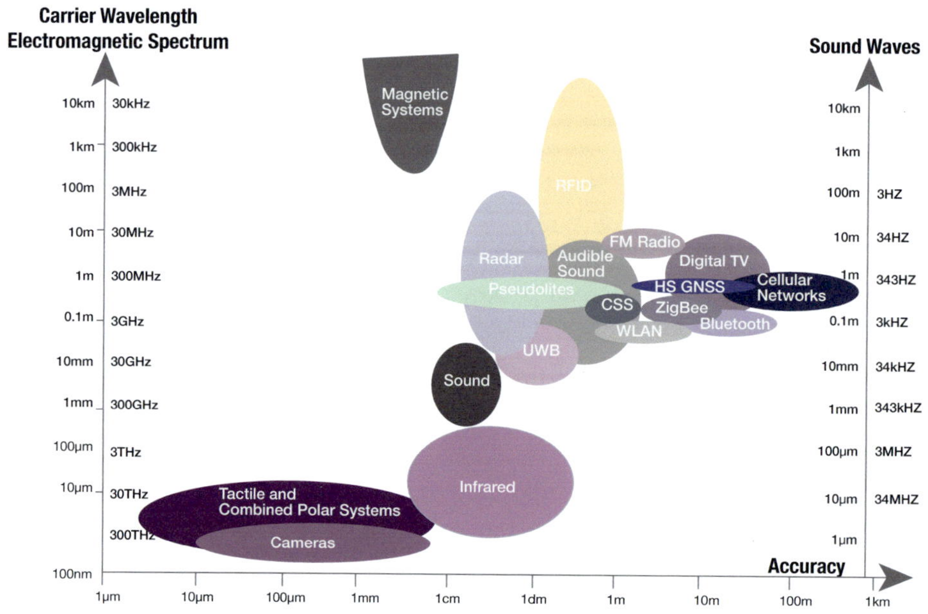

Source: [121]

Alternative location accuracy improvements include a method that was reported in a recent issue of GPS World. A team from the Radionavigation Lab collaborated with researchers from the University of Texas, Austin, and used smartphone cameras to improve GPS location to centimetre accuracy by quickly building a globally referenced 3D map of the surroundings. A special receiver (GRID) was used to extract carrier phase measurements from the low-cost antennas found in mobile devices. The method has a multitude of potential applications such as traffic collision avoidance, accurate RPAS parcel delivery and the outdoors use of VR-headsets [122].

LBS (Location-Based Services) use location-enabled devices to track the users and to determine the visitor's use-pattern. Google introduced 'visit monitoring', a feature that collects information about places visited in the day-to-day life of users [123], and developers are working on innovative uses of this information. Certainly, mobile advertising will be a driving force behind the research. It was found that mobile ads create vastly superior sales results. Some are yielding an average of US$30 in sales per 1000 impressions in contrast with only US$13 on a desktop, and resulting in an average return on ad spend of 257%. This result came from a study of 83 mobile campaigns with an average duration of 11 weeks, in seven CP categories [124].

In 2016 advertisers in the USA are expected to more than double their spending to over US$40 billion in order to reach mobile users (on phones and tablets). This move follows the strong trend in using mobile devices, which was on a par with desktop and laptop usage time in 2014 [125]. According to a Berg Insight report, LBS advertising in 2018 will be 2% of total global advertising expenditure for all media, and 7% of all digital advertising, with BLE (Bluetooth Low Energy) beacons being disruptive game changers. LBS is predicted to grow with a CAGR of 54% from US$1.66 billion in 2013 to US$14.8 billion in 2018 (Eddy, 2014) [126]. Moreover, LBS allow much more targeted advertising campaigns. McDonalds in the Netherlands saw a 700% increase in offer redemption using location-based advertising on the VMob platform [20].

LBS can be segmented into a range of markets including: indoor; LBS search and advertising, LBS tracking, LBS navigation, LBS infotainment, analytics, recreation and fitness. The key LBS vendors are mentioned in the footnote[6]. This information is included in a comprehensive report on the Global LBS markets (2015–2019) by Technavio [127]. ABI research predicts that 400 million BLE beacons will be shipped by 2020 [128], with many large-scale installations (comprising thousands of beacons) already completed.

Apple has developed indoor navigation and positioning with its iBeacon through the release of its iPhone 6 and iOS8. The M8 motion co-processor includes a barometer on top of the compass, gyroscope and accelerometer. The barometer senses the air pressure to the degree of accuracy that the phone can be located with respect to floor levels [129]. Apple acquired the company WifiSLAM with technology that gives 2.5m accuracy by using only ambient Wi-Fi signals after initial fingerprinting of a building with a smartphone. Apple's partnership with TomTom, which in turn has a deal with Micello (a major indoor map provider), is another stepping stone in the race to make ubiquitous indoor navigation a reality [129].

---

[6]Adnear; Alibaba; Facebook; Foursquare; HERE; Telecommunication systems; TomTom ; AdMoove; Aisle411; Android; Bafang; Bing Maps; Blinksight; Bouncity; Celcom; CEVA; Cisco; Clingle; CommScope; CompassCom; DbyDx software; Deveryware; Dialme 24; Earthlink; Ekahua; Ericson; ESRI; Fastmall.com; Field Power; FindMap; Genasys; IndoorAtlas; IndoorGo; Indosat; Insiteo; Intellinet; Invisitrack; IQService's; Jatis Mobile; Komli mobile; Lighthouse signal systems; Link Communication; Locaid; LociLoci; Locotronix; LoJack; Mappoint Asia; MetaCarta; Navizon; NetEase; NextNav; Omnisense; Point Inside; Quuppa; Roger wireless; Sensewhere; Skyhook; SmartTrust; Taximonger; Telenity; Telesonera; TelideoTrimble; TruePosition; Trusted Position; Ubisense; Useful networks; WiseSec; ZOS.

These developments form the basis for new location infrastructures and are essential enablers for the Internet of Things, including giving everyday objects location-smarts.

1.6 RPAS (REMOTELY PILOTED AERIAL SYSTEMS)

ABI Research estimates the small RPAS market will surpass US$8.4 billion by 2018. The commercial RPAS sector alone has an estimated 51% CAGR, 2014–2019, with revenues exceeding US$5.1 billion by 2019. On the other hand, BI Intelligence estimates a growth in the civilian drone market of 19% CAGR and 5% CAGR in the military market between 2015 and 2020. They see the most notable RPAS companies to be Sensefly (owned by Parrot), Aeryon, CybAero, DJI and Gryphon [130].

Key industry drivers will not be primarily hardware and platforms, but industry-specific applications and data as well as operator and modelling services (see Figure 45) [131].

According to Dan Kara, Practice Director of Robotics at ABI Research:

> The commercial sector is the sweet spot for the sUAS[7] market, a fact recognized by both defence industry suppliers such as Elbit Technologies, AeroVironment, and Aeryon Labs, as well as providers to the Prosumer/Hobbyist marketplace including DJI, Parrot, SenseFly, 3D Robotics, and others. As a result, both groups of sUAS makers, along with other classes of solution providers, are aggressively targeting the commercial sector through acquisitions, internal development, partnerships and investment. [131]

FIGURE 45: GLOBAL AERIAL DRONE MARKET, 2013–2024

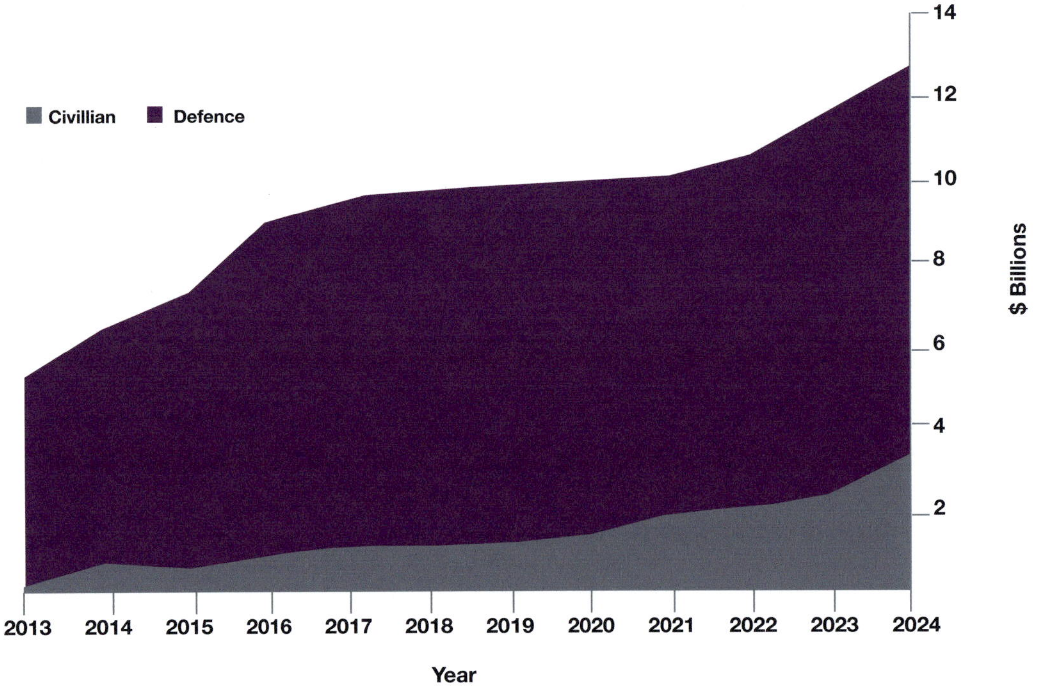

Sources: Teal Group; BI Intelligence Estimates, Michael Toscano

Source: [130]

[7] sUAS; small Unmanned Aerial Systems. Also known as RPAS.

Figure 46 shows the investment in RPAS Research and Development by country.

FIGURE 46: RPAS R&D BY COUNTRY

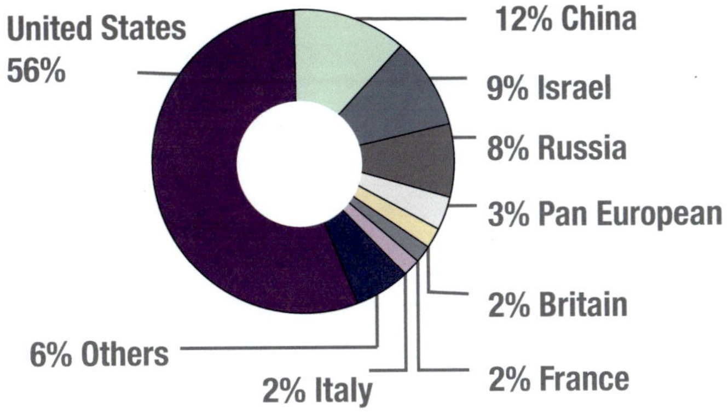

Source: [132]

Figure 47 shows a subset of an overview map of the Civilian RPAS.

FIGURE 47: DRONE LANDSCAPE, 2014

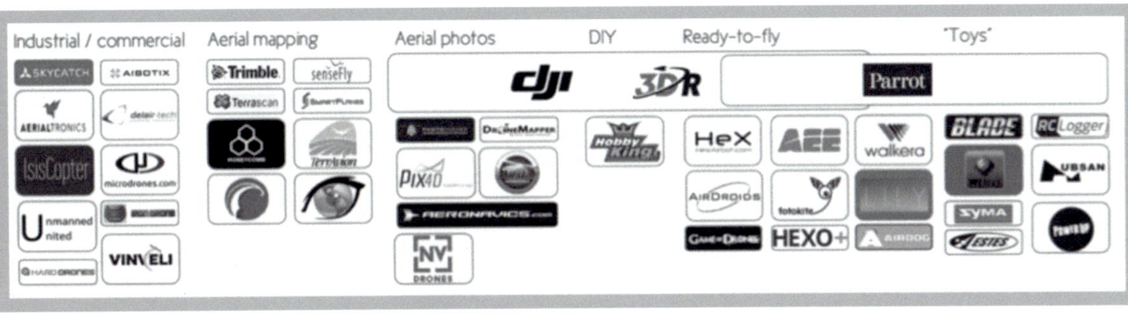

Source: [133]

A very active RPAS community has developed including one company called dyidrones.com. RPAS have many uses, such as taking selfies – 'dronies' like Nixie (winner of MIW challenge) [24] or delivering orders such as Google's Project Wing, a drone program of Google X Lab, in August 2014. This drone has vertical take-off and landing capability and delivers goods via a pulley system [134].

Two students from the University of Wollongong have developed a life-saver quadcopter that can deliver flotation devices to distressed swimmers. The flotation device is carried by magnets on a RPAS and released remotely by an onshore lifesaver [135]. In December 2014, students of Shahjalal University mapped environmental damage to mangrove forest after a 350,000 litre oil spill, due to a capsized tanker, close to a World Heritage site between Bangladesh and India [136].

There is much merit in utilising RPASs for dangerous jobs (e.g. roofs, transmission power, wind turbine inspections or the mapping of collapsed buildings in emergency situations). Venture capitalists are now interested in funding RPAS companies, and large geospatial firms such as Woolpert and Merrick are offering RPAS services [137].

Small RPASs are readily available for purchase from shops and online, yet it is often unclear where and when they can be flown legally. The same RPAS flown recreationally on the weekend is subject to different laws when operated in a professional income-earning context [138].

The APPS RPAS logger and Airmap Beta are amongst the emerging APPs that help users to find spaces where drones can be flown legally. The integration of small RPASs with the existing air-traffic control systems also still needs to be worked out [139] [140]. China-manufactured DJI Drones have added geo-fenced 'no-fly zones' and now block drones from operating in the vicinity of 350 airports around the world in an effort to minimise collisions between drones and manned aircraft [141]. The Civil Aviation Safety Authority (CASA) has also requested that written guidelines accompany the DJI drones, after incidences such as crime-scene recording and the endangerment of emergency airborne vehicles during bushfires [142].

In 2002 Australia introduced the world's first RPAS regulations; the creation of which caused early R&D and international investment.[8] At the time, Australia developed several world-first technologies, some of which are now licensed globally. Recent attempts to update CASR101 have again been very slow, however on 29 September 2016 some red tape was removed for RPAS operators flying crafts under 2kg [143]. The Airservices operational RPAS concept guidelines were released in May 2016 [144].

While the changes will certainly benefit small RPAS users [145], Australia no longer enjoys the benefit it once had of being 'first in the field'. Due to the reduced rate of Australian reform in years leading up to 2016, indigenous RPAS R&D opportunities and investments have fallen behind other parts of the world. The push for change has so far been from the bottom up, from R&D organisations, industry associations and RPAS operators.

Now with the introduction of the OneSky [146] system for integrated civil and military-piloted aircraft operations, Australia would also benefit from the integration of RPAS in civil airspace.

Six key recommendations were made in the Australian "Eyes in the Sky" [147] covering peak-body consultation, privacy concerns, the use of surveillance equipment and air safety. Another review is to be conducted in 2016 due to the rapid development of the technology.

Flying cars (Terrafugia, Figure 48) and hover-motorbikes (Aero-X, Figure 49) can now be pre-ordered, the latter being delivered in 2017 for a price of US$85,000 [148]. Another hoverbike project, by Australian Chris Malloy, was to be funded by Kickstarter but he now works with the US Army Research Facility to further develop his hoverbike [149]. Chinese drone company Ehang presented an autonomous personal transport drone at CES 2106 (see Figure 50). All will need major legal reforms to become 'road'/air legal.

FIGURE 48: TERRAFUGIA FLYING CAR

Source: [150]

FIGURE 49: AERO-X HOVERBIKE

Source: [148]

[8] Personal communication from George Curran, Business Development CRCSI, February 2016.

FIGURE: 50: EHANG AUTONOMOUS FLYING VEHICLE

Source: [151]

FIGURE 51: AIRMULE FLIGHT

Source: [152]

Air Mule, a two-person VTOL entirely autonomous flying ambulance can transport over 1000 pounds over more than 30 miles. The vehicle completed an untethered 2.5-minute maiden flight in December 2015 (see Figure 51) [152].

2. SMART SYSTEMS

2.1 SPATIAL DATA

Governments have realised the value of open data and are actively pursuing ways to enable easy access for all. Tim Berners-Lee, often referred to as the inventor of the World Wide Web, is President and Co-Founder of the Open Data Institute. This is a UK-government backed (£10m) initiative linking universities and commercial supporters that advocate access to open data [153].

FIGURE 52: LANDSAT DOWNLOADS SINCE OPEN-DATA POLICY

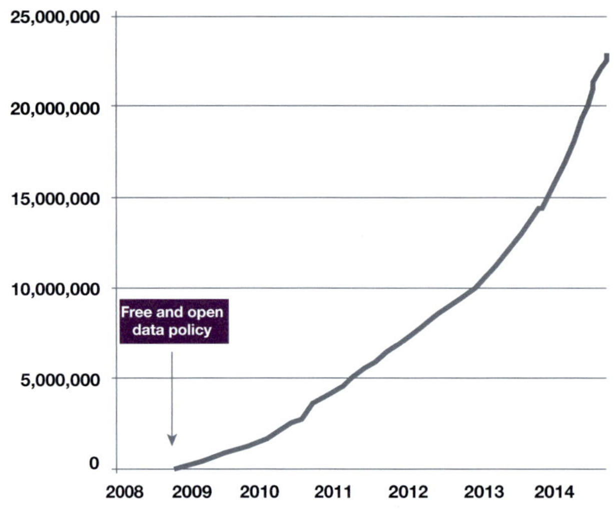

Landsat scenes downloaded since the advent of the public good data policy (Credit: USGS)

Source: [154]

The open data policy has been applied to Landsat satellite data, and consequently data were made freely available. Since then a significant increase in user downloads can be observed (Figure 52). Australia does not own any EOS satellites and depends on access to data from satellites controlled by overseas organisations. The paper "Risks of Data supply of Earth Observation from Space for Australia"[154] notes:

- The nature of the best available public good data streams for the four main instrument types will change. Some of these changes will be positive, and others, negative; but overall, supply for these types is expected to remain strong.
- The additional data streams coming online in the 2015–2016 timeframe will greatly increase data volumes. This is due to greater spatial and spectral resolution, to more supply sources and more accessible archives, and because of a greater revisit frequency. These factors mean that ground segment and data management solutions such as the Australian Geoscience Data Cube and Google Earth Engine, will likely become more important in enabling users to manage and make use of this new data. This likelihood also underscores the need for strong coordination with international partners, and potentially for new coordination frameworks with CEOS Virtual Constellations, for example.
- Advances in small satellite technology have cut the cost of investing in a specific satellite space segment. While public-good data streams from overseas will continue to be the dominant supply, the business case for an Australian national Earth-observing satellite or satellites, or contribution to an international partnership, looks increasingly attractive. Space-segment investment would need to be supported by the ground segment, which would entail additional requirements to receive, standardise, and apply data in a common framework such as the Australian Geoscience Data Cube.

In June 2015 the Commonwealth Scientific and Industrial Research Organisation (CSIRO) signed an agreement with the European Space Agency (ESA) that will give Australia better access to data from Europe's satellites; the Sentinel satellite constellation in particular [155]. Australia in return will cooperate with Europe by sharing its scientific expertise. With over 100 government programs and hundreds of researchers, Australia is globally one of the largest users of Earth Observation data. It is therefore important to adequately plan for sufficient funding to build robust data infrastructures in the country to handle upcoming rapid increase in spatial data.

The Department of the Prime Minister and Cabinet in Australia has given a high priority to making public sector data available and has published a road map to that end (see Figure 53) [156]:

FIGURE 53: PUBLIC SECTOR DATA ROADMAP

Recommendations: Pulic sector data roadmap

It is recommended that a clear mandate and several policy projects be initiated to generate confidence in public sector data over the short term and structural changes be implemented over the longer term to transform the way the APS uses data.

1-6 Months:
Build confidence and momentum

1. **Secretary of PM&C signals data is a priority** by promoting the agenda across the APS
2. **Commission several high value projects** focused on key policy questions, proving the value of public sector data and removing barriers
3. **Build external partnerships** to foster demand and encourage the use of public sector data
4. **Agencies publish readily available datasets** on or through data.gov.au and schedual future releases
5. **Build data analytics capability** by bolstering existing efforts, partnering externally and investing in pockets of excellence
6. **PM&C coordinates progress** on demonstration projects, develops APS data policies and takes steps to rationalise governance arrangements through cross-agency team (see recommendations 7-8)

1-18 Months:
Systemstise the use and release of public sector data

7. **implement a data policy and governance framework** that includes:
   a. **a public policy statement** that could be delivered by the PM outlining the narrative and high-level principles
   b. **a simple governance model for data policy** that incentivises and drives consistant transformation through clear ownership, links to external groups, a high level steering group, partnerships with the private sector and data champions within agencies
   c. **a requirement for evidence based policy** development to motivate data use in policy.
8. **Build and maintain trust,** engage with the public to understand the benefits to citizens and address privacy concerns on public sector data
9. **Establish a trusted-acess model for sharing integrated data** across agencies
10. **Create and publish a searchable whole-of-government data catalogue** to make government data holdings discoverable
11. **Develop a Commonwealth Government high-value dataset framework** to inform prioritisation and release
12. **Publish data management standards** to simpllify the processes of integrating data and publishing on data.gov.au
13. **Establish a consistent and transparent approach** to user charging to provide clarity on how to manage costs
14. **Create a legislative environment that supports the use of data while maintaining privacy** by initially providing guidance and ultimately revisiting legislation through a possible Productivity Commission inquiry
15. **Promote innovation in public adminstration** through targeted awards and resources

Source: [157]

The demand for data in Australia could be categorised in the following sectors (Figure 54):

FIGURE 54: TOP 5 REQUESTED DATA THEMES IN AUSTRALIA

Top 5 data themes	Research sector: Top 5 most requested data	Private sector: Top 5 most requested data
1 Spatial and land	1 Census and Labour Force Longitudinal Data	1 Geo-coded National Address File
2 Socio-economic	2 Geo-coded National Address File	2 Cadastre and Land Tenure
3 Health	3 Personal Income Tax and Business Tax	3 Geo-coded Australian Business Register
4 Transport	4 Health Benefits Data (linked MBS/PBS)	4 Health Benefits Data (linked MBS/PBS)
5 Environment (incl energy)	5 Social Security Payments	5 Census and Labour Force Survey Data

Source: [156]

Data can be viewed on the National Map, a central website that gives access to spatial data from Australian government agencies. The software underpinning the system has been developed by project lead Data61 in cooperation with other government agencies [158]; and additional open-source software used include TerriaJS, Cesium and Leaflet. The portal facilitates easy, open and direct access to data held by a variety of government agencies, although it does not hold the data itself.

The use of open data is popular, as can be seen by the data requests submitted to sites hosting the Queensland Globe (with about 500 datasets) (see Figure 55). Note the increase in data requests from mobile devices shown in Figure 56.

FIGURE 55: DATA REQUESTS TO QLD GLOBE

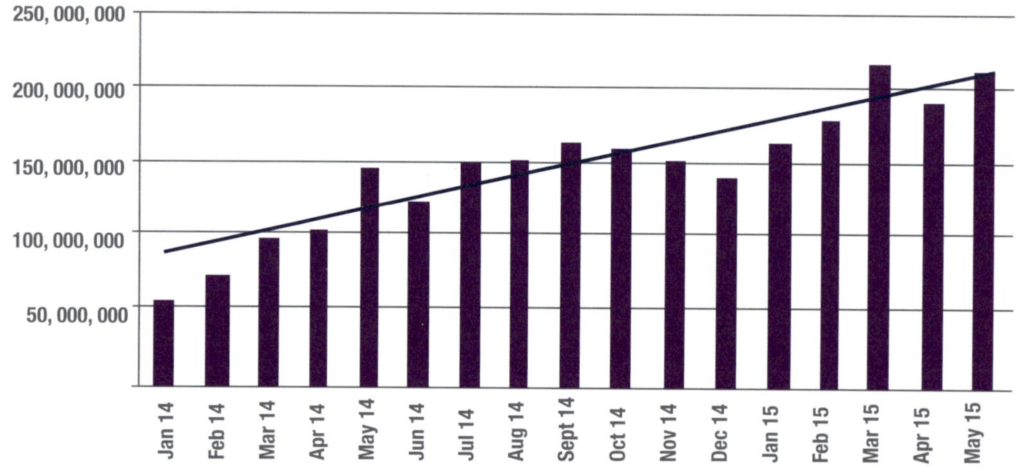

Source: [9]

[9] Personal communication from Steve Jacoby Executive Director, Land & Spatial Information Division, | Department of Natural Resources and Mines, 17 June 2015.

FIGURE 56: DATA REQUESTS FROM MOBILE DEVICES

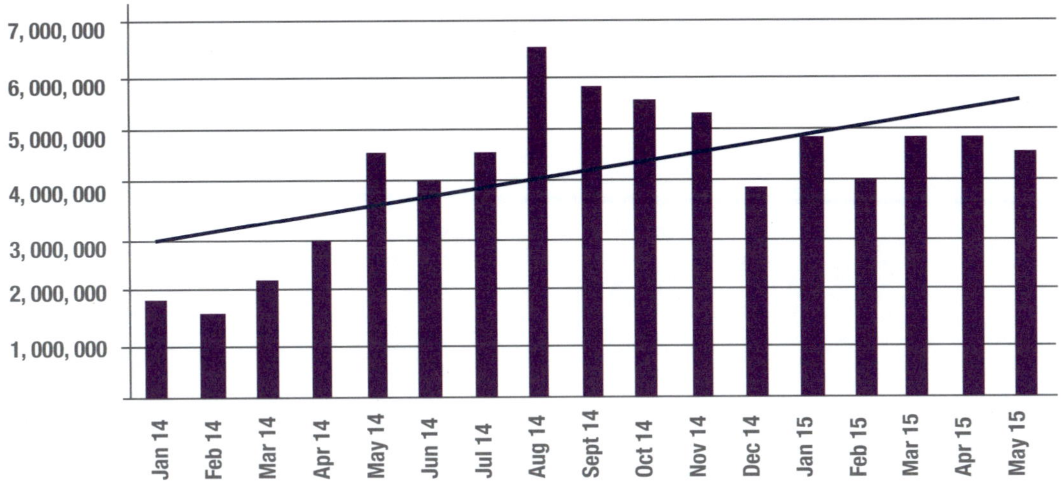

Source: [10]

The New South Wales Globe data contain: (see Figure 57)[11]

- Imagery: medium- and high-resolution aerial; satellite, both governmental and privately sourced
- Response imagery for flood, fire and storm, for example
- Terrain data: 25m DTMS, 5m DTMS, LiDAR and SRTM at various locations
- Places, such as. population centres and suburbs
- Boundaries, for example, coastlines, local government, localities, electoral states and federal
- Roads, such as arterial and minor roads
- Rail: railway lines and railway stations, for instance
- Addresses, such as property and address
- Land parcels: lot and deposited plan, property boundary, labels, for example
- Surveying: survey marks and CORSnet stations, for instance
- History, such as 1943 Sydney imagery, historical maps
- Property sales information
- Land values: free access to 14 years' historical data
- Australian Bureau of Statistics (ABS) Census data

[10] Personal communication from Steve Jacoby Executive Director, Land & Spatial Information Division, | Department of Natural Resources and Mines, 17 June 2015.

[11] Personal communication from Belinda Jackson, Land & Property Information | Department of Finance, Services & Innovation, 26 October 2015.

FIGURE 57: NUMBER OF HITS TO NSW GLOBE

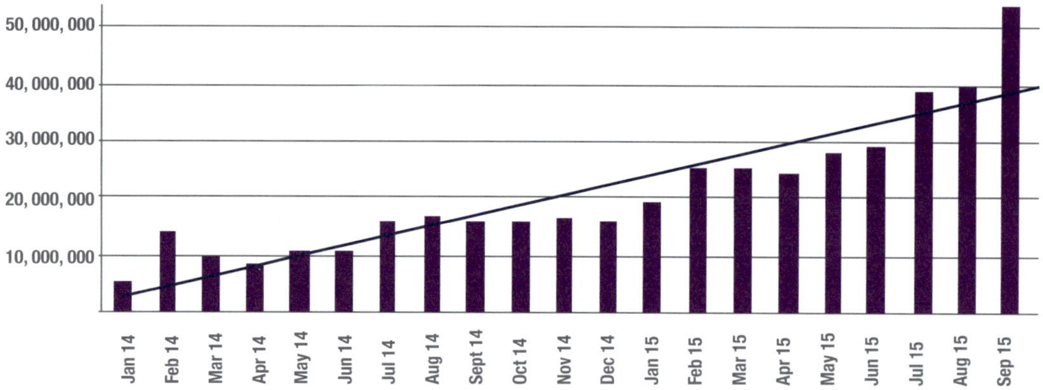

Source: [12]

Victoria has 628 spatial datasets available in the Victorian Government Data Directory [159]. VicRoad opened its data and has been named a role model for other open-data initiatives [160].

Landgate **(Western Australia)** reported[13] 3153 datasets on their website, supplied via WMS (Web Map Service) and WFS (Web Feature Service). The data downloaded totalled 13.9851 GB (from 1 January 2014 to 30 June 2015). Data requests doubled from 20 million in July 2015 to almost 40 million in October that year. For the reporting period from 1 January 2014 to 30 June 2015 the data class types were as follows:

Class Type	Gigabytes Out
Public	3,526.45
Subscription	10,473.55
Restricted	334.70

VIRTUAL GLOBES

The role of globes was foreshadowed by Al Gore in his 1998 speech on virtual globes predicting the worlds citizens would one day be able to interact with virtual globes depicting scientific and cultural information [161]. Building on the concept of virtual globes, the International Society for Digital Earth was formed in 2006 in Beijing to foster the vision of a Digital Earth.

During the G20 meeting in Brisbane, Australia in November 2014, the potential for open and free access to spatial information was demonstrated through the G20 Globe [162] [163].

[12] Personal communication from Belinda Jackson, Land & Property Information | Department of Finance, Services & Innovation.
[13] Personal communication from Melissah Johnston, Media and Outreach Officer, WALIS, November 2015.

A review paper, commissioned by the Open Digital Earth Foundation evaluates the Digital Earth globe technologies currently available. These include case studies [164] such as the Google Earth and Bing Maps 3D Platform, Cesium (AGI), Bhuvan 3D (India), QLD Cube Globe, NASA World Wind, WebGL Earth 2 (Klokan Technologies), Marble (KDE), Skyline Globe, ESRI ArcGlobe, EV Globe, SuperMap GIS, Digital Earth Science Platform, and osgEarth [164]. The review summarises 21 globes operating off 16 unique globe architectures, of which 6 were fully open whilst 8 were operating from proprietary architecture or with some form of restricted access or pricing.

2.2 SPATIAL DATA INFRASTRUCTURE AND ANALYTICS

For the last two decades the spatial community has been working on the concept of harmonised spatial data through a common infrastructure. Data producers, aggregators and end users within the infrastructure are able to discover, interact, analyse and seamlessly develop reports; irrespective of varying spatial data quality, of heterogeneous data formats or of the different data models. These options are available on any device, and at any time and place. During the first decade of its research, the spatial community focused on creating frameworks; data licensing models; common data models; the interoperability of data and data transfer standards. In the second decade, the community leveraged the power of internet; smartphones; web-based data marts/portals; crowdsourcing and web services. Today the main focus is on providing geospatial data in the form of distributed spatial web services, data retrieval through catalogues, and visualisation in the form of Web Map Services (WMS) [165].

It has emerged that the Spatial Data Infrastructure (SDI) work has been of more interest to governmental and academic sectors than to the private sector. Governments have heavily invested in the work to ensure that the spatial data created using taxpayers' money is presented to citizens in a useful way; recently with free access to spatial information (SI) through open-data policy and open government. The INSPIRE initiative from the European Union, The USSDI in the United States and Canterbury SDI in New Zealand are some good examples of regional, national and local government SDIs, respectively. The advent of Smart Cities by Urban Planners, 3D-city models, Building Information Models from the construction industry and architecture community has also recently provided an opportunity for the SDI community to develop the concepts of Indoor SDI.

SEMANTIC WEB AND SPATIAL DATA INFRASTRUCTURE

Powered by Semantic Web technologies, the Linked Data paradigm aims at weaving a globally interconnected graph of raw data that transforms the ways we publish, retrieve, share, reuse, and integrate data from a variety of distributed and heterogeneous sources [166]. In practice, however, this vision faces substantial challenges with respect to data quality, coverage and longevity, the amount of background knowledge required to query distant data, the reproducibility of query results and their derived (scientific) findings and the lack of computational capabilities required for many tasks. Research on geo-semantics plays an increasing role in supporting complex queries and retrieval across heterogeneous information sources as well as for service orchestration, semantic translation and on-the-fly integration [167]. Resource Description File (RDF), which is a standard model for data interchange on the Web, facilitates data merging even if the underlying schemas differ, and it specifically supports the evolution of schemas over time without requiring all the data consumers to be changed [168]. The OGC GeoSPARQL standard supports representing and querying geospatial data on the Semantic Web. GeoSPARQL defines a vocabulary for representing geospatial data in RDF, and defines an extension to the SPARQL query language for processing geospatial data. In addition, GeoSPARQL is designed to accommodate systems based on qualitative spatial reasoning and systems based on quantitative spatial computations [169]. With the advent of RDF, combined with GeoSPARQ, the end users are able to utilise the semantic web platform to interact with the spatial web services published by producers of spatial data, mainly mapping agencies.

NEXT GENERATION SPATIAL DATA INFRASTRUCTURE

The next generation spatial infrastructures must address multiple contemporaneous issues within the spatial data supply chains (SDSC) [170].

The CRCSI in Australia has set up an exclusive program to examine technical solutions to the spatial industry supply chain problems through the application of semantic web and linked data technologies. The program will address the boundaries and gaps identified to bring about seamless supply chain integration and operation. The research aims to establish a universal approach that can be framed to deal with SI supply chain issues that allow for the understanding and automation of the supply chain process that incorporates multiple data sources, including crowdsourced data. The research is undertaken with the following components of the supply chain:

- User driven Pull method for data consumption against the traditional Push method from data producers
- Automatic query-driven orchestration
- On the fly data federation
- Dataset conflation
- Provenance, trust and quality
- Intelligent geographic search

The research projects have arisen from a consultation process that examines spatial-data supply chains within Australia and New Zealand. Major partners of the research program are from respective State and Federal government departments in addition to some private sector partners.

Anticipated benefits:

- Improved efficiency in data curation and provision
- Cost savings through automation
- Broader access of complex spatial data to end users
- Smart data conflation process
- Improved data provisioning to end users
- Ease of geometric data retrieval by novice end users

OPEN SPATIAL ANALYTICS

Globally, the open data agenda and policy of governments has resulted in the availability, both to businesses and the public, of complex spatial data of varying quality. Voluminous spatial data is available through government portals. However, analysis of the information provided through these data portals is limited: simple spatial analysis tools are only available to the user through the portals, and the end users are often required to perform complex spatial queries in order to resolve a business or an end user problem. The next challenge is to provide the means of performing real-time analysis through spatial data portals, including the currency of the spatial data available through open data initiatives.

To facilitate the spatial analysis of data, end users would benefit from a virtual spatial analytic workbench to use through a web platform. The workbench could host open geoprocessing/analytic tools that end users can customise to interact with the open spatial data in order to generate rapid outputs. The open spatial analytics tools could be managed through the creative commons licence process, with the user community maintaining the tools and versioning. The advent of open source statistical software such as 'R', coupled with spatial analysis processes may provide a powerful geostatistical analysis platform for end users to work with the open data.

The CRCSI in Australia has recently developed an 18-month research program, entitled, "Rapid Spatial Analytics (RSA)" that aims to rapidly create and value-add SI products, manually or through automation, both from devices and cloud-based infrastructure. The RSA Program revolves around five carefully selected themes: Spatial Workflows, Real-Time Spatial Analytics, Big Spatial Analytics, Foundational Spatial Analytics, and Visualisation and Decision-Making. The term "Rapid", which is present in every theme, is a concept that has different meanings for each area of application. Rapid can mean "near real-time provision of outputs" (Real-Time Spatial Analytics), "efficient processing" (Big Spatial Analytics) or "fast access and integration of data" (Foundational Spatial Analytics).

2.3 MAPPING SYSTEMS

The convergence between CAD, BIM and GIS continues as there is a need for greater interoperability, broader access, and data and tools integration [171]. Several groups are working on this integration (i.e. ISO BIM-GIS Ad Hoc Group, OGC-building SMART alliance partnership and InfraBIM project) [172].

Autodesk has added analytical (GIS-like) capabilities for road-building projects or wastewater modelling, and compute-intensive operations are executed in the cloud. InfraWorks 360 mobile APP visualises BIMs on the job site; Autodesk Showcase brings 3D CAD models alive for interactive presentations [173]. Users can manage field processes such as quality, safety and checklists, as well as distribution of and access to plans and drawings.

Autodesk and Leica Geosystems are collaborating to streamline construction workflows. The collaboration involves reality capture and BIM to field layout solutions with laser scanning, building construction layout and similar markets [174].

Bentley now also combines BIM and GIS with its recent purchase of SiteOps. It combines design and cost models for optimised workflow and helps engineers and developers understand site preparation costs as a function of design [173].

2.4 BIG DATA, ALGORITHMS, APPLICATIONS

The power of big data in the cloud and access to powerful processors is immense – it means gaining knowledge with relative ease compared to any process just a few years ago. The following examples illustrate the power of big data collection and analysis, and the derived applications.

Geoscience Australia scientists have unlocked 25 years of archived satellite imagery with their Data Cube, using the high-performance computers of the Australian National University's National Computing Infrastructure (NCI), to monitor land and water resources and provide unique information on flood risk and ecosystem management [175].

FIGURE 58: 25 YEARS OF WATER OBSERVATION FROM SPACE

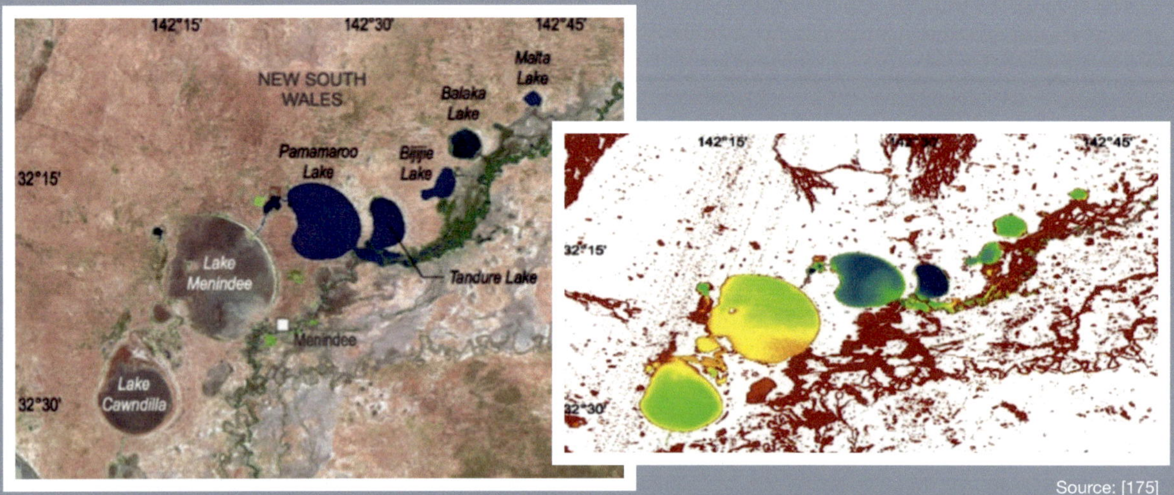

Source: [175]

Figure 58 shows analysed satellite imagery over 25 years. The permanency of hydrological features in the landscape are colour coded; red depicts areas that hold water 1% of the time; green areas, 20% of the time; while blue areas always hold water (100%).

Mobile network data have the potential to restrain infectious diseases such as Ebola. Researchers used data from the Orange Telecom network in Cote d'Ivoire, Senegal to map the aggregated call patterns of one million anonymous phone users at transportation hubs in order to understand the potential routes of virus spread within the country [176]. The US Department of Defense has modified the Constellation system (a WMD bio-surveillance prototype) for contact tracing people with direct contacts to sick Ebola patients, and for using this information in outbreak responses.

CSIRO, in partnership with Intel, is also developing algorithms for the typical behaviour of honeybees to gain a better understanding of what is harming bees in an effort to secure crop pollination. The bees were monitored with sub-millimetre sensors and vast amounts of movement data was modelled in a GIS [177]. Both typical and atypical behaviour could be documented.

Additionally, a team at the University of New South Wales analysed the first available differential interferometric SAR images after the April 2015 quake of 7.8 magnitude in Nepal, and realised that another major earthquake would hit Nepal soon after. In before-and-after quake imagery they recognised a displacement of 1.5m; however, the modelled displacement should have been 3.2 metres. The team therefore warned the authorities to prepare for disruptions to the recovery efforts. Ten days later another quake, this time of 7.3 magnitude, hit Nepal [178].

At Kent State University, researchers are working with police to get a more complete picture of crime. Officers, while driving through the neighbourhoods of Akron, record geo-tagged video footage, and dictate information for specific points of interest such as a vacant property known as a drug house, for example. These geo-narratives are brought together with traditional crime data for more comprehensive insights [179]. Likewise, the body-worn cameras of field officers have proven to be an invaluable tool for recording and verifying eyewitness statements at the crime scene and in the court room [180]. In Victoria, police officers proposed to shoot fleeing cars with a GPS tracker rather than risking lives in high-speed pursuits [181]. The company Starchase already offers such a car-mounted GPS tracking gun to police, and the system has been tested both by Iowa and Florida police [182].

2.5 SMART SYSTEM APPLICATIONS

The integrated global Internet of Things is feeding big data, which in turn will use advanced analytics to create actionable intelligence with predictive algorithms to program automated SMART systems with improved efficiencies. Cameras, scanners and sensors will provide data to smart systems in every sphere of life: cities, energy, transport, agriculture, natural resource management and health etc. will all benefit from integrated smart systems. Sensor data and smart algorithms will merge to create the most futuristic applications.

Smart systems have the possibility of transforming whole supply chains, such as the beef supply chain, for example.

Millions of pounds funding has been awarded to Silent Herdsman, a scheme allowing Scottish farmers to track the health of their herd through internet-enabled sensors in the cow's electronic collar; allowing for increased milk productivity and the early identification of potential bovine health issues. The firm now plans to roll out the application across the globe [183]. Silent herdsman detects heat in cows, significantly enhancing calving rates. Moonitor [184] also monitors cow behaviour through a solar-charged collar, which is a sensor network (with data upload via satellites [185]) that can alert farmers of weight gain, oestrus events, pregnancies and other medical conditions. Cattlewatch is an alternative system for monitoring the herd, promising a yield increase of 10–20% as well as farm-cost reductions [186].

The Precision Pastoral Management Systems (PPMS) [187], developed by the Cooperative Research Centre for Remote Economic Participation, incorporates remote livestock management (automated weighing and drafting); telemetry, which involves satellite observation of land and pasture; water management and cloud-based analytics. The PPMS package saves farm managers labour, time and money, improves livestock productivity, increases sustainability, and protects both vegetation and wildlife in arid rangelands and savannahs.

In his paper “Tracking and traceability system using livestock iris code in the beef supply chain”, Zhao [188] reports on cattle iris scans. Iris scanning can now be done from smartphones [189]; and theoretically, smartphone cameras could be mounted to Kopter RPASs for in-field animal recognition.

Smart sensor tattoos cause the noses of stressed animals to glow blue so the farm manager can step in to help reduce stress levels in cattle [190] (in particular before slaughter to ensure the best produce quality).

The NRM Spatial HUB, developed by the CRCSI, helps graziers optimise infrastructure developments as well as maintain sustainable grazing regimes. Information is extracted from satellite imagery and served in online property-maps that graziers can easily interact with, as seen in Figure 59–62. Baseline Landsat imagery covering the last 30 years can be analysed for biomass in the matter of a few seconds and the results delivered to graziers in near real time [191].

FIGURE 59: THE OPPIS USER INTERFACE SHOWING A TIME-SERIES FRACTIONAL COVER GRAPH FOR A USER-DEFINED REPORTING AREA

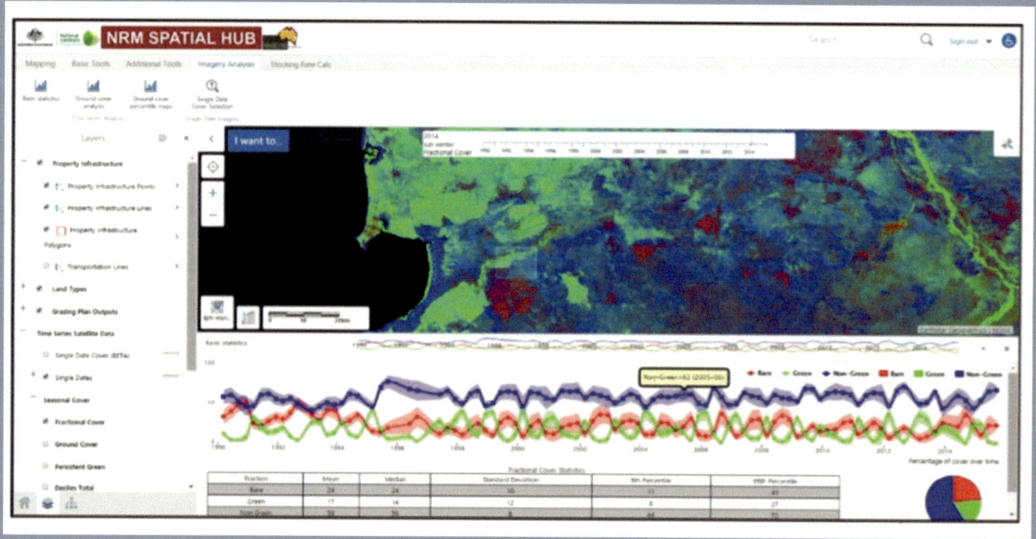

Source: [191]

FIGURE 60: TIME-SERIES OF GROUND COVER DATA, GRAZING CIRCLES AND ANALYSIS

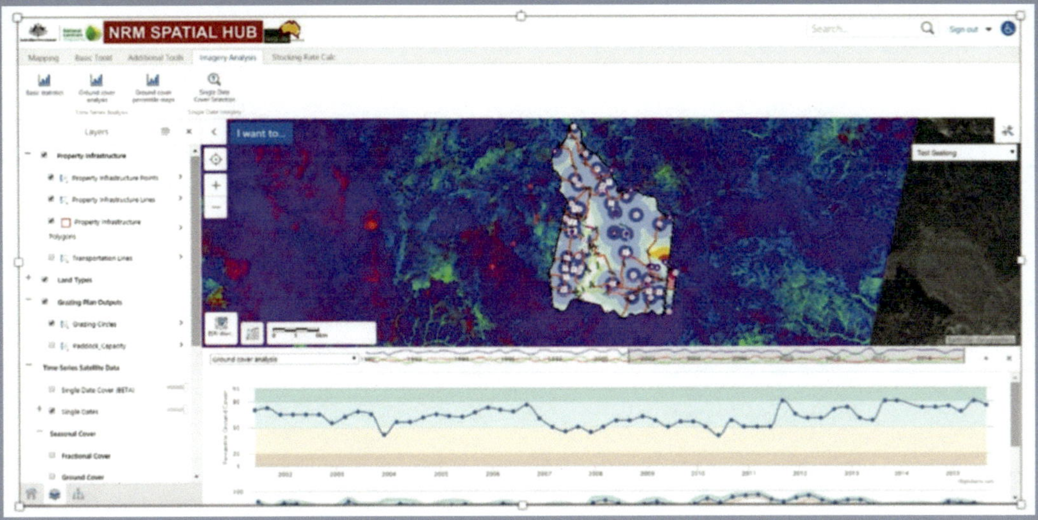

Source: [191]

FIGURE 61: GRAZING CIRCLE ANALYSIS SHOWING PERENNIAL (BLUE), EPHEMERAL (ORANGE) AND ESTIMATED PADDOCK UTILISATION

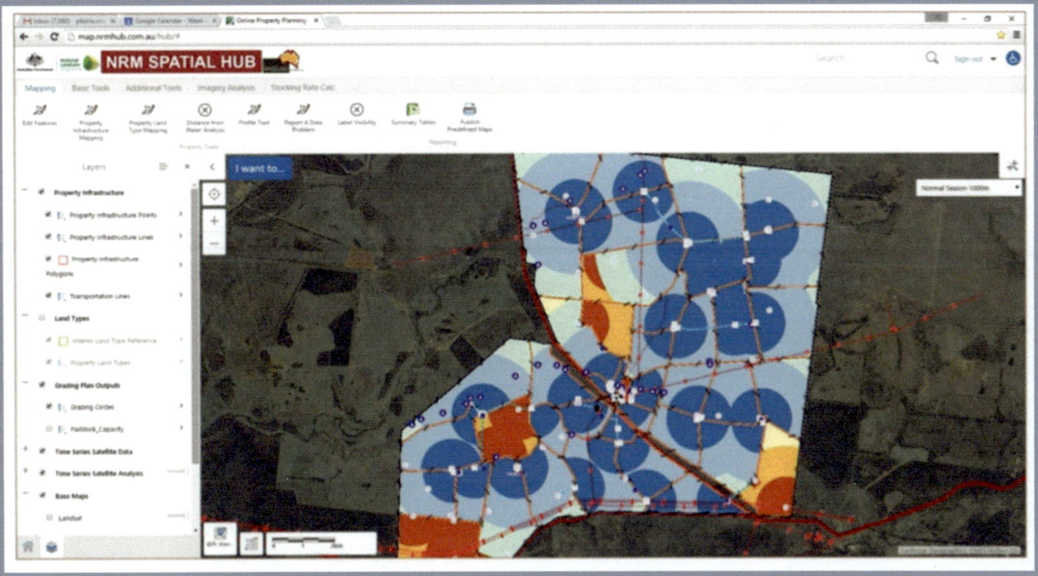

Source: [191]

FIGURE 62: OUTPUT MAP FROM THE SAFE-CARRYING CAPACITY CALCULATOR

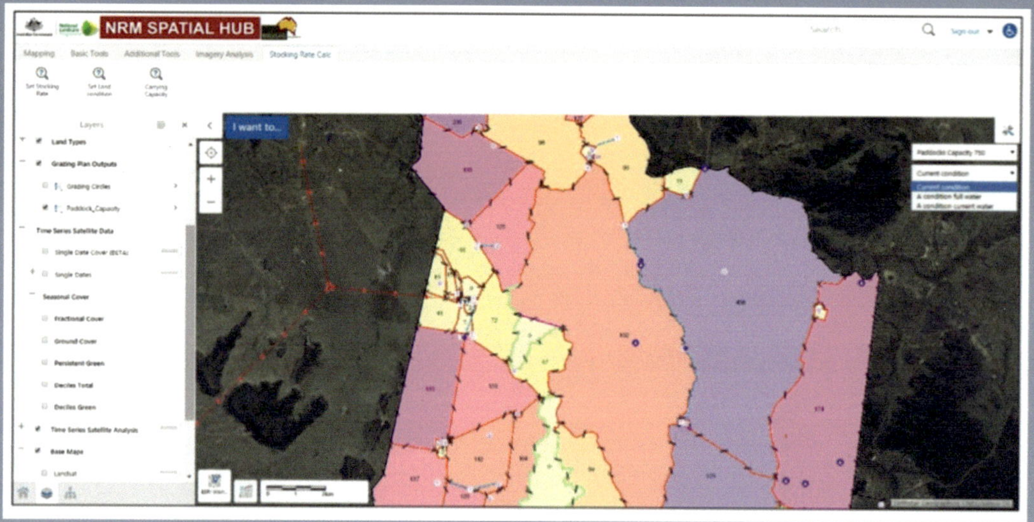

Source: [191]

Optimised grazing regimes may be enforcable with virtual fencing where farm animals are trained to stay within the boundaries using wearable GPS-location tracking and audio or other cuing using predetermined locations [192].

To optimise overall farm performance, John Deere and AGCO are beginning to connect farm machinery as well as irrigation systems, soil and nutrient information and weather reports with crop and commodities futures [193].

FIGURE 63: SYSTEM OF SYSTEM SOLUTIONS FOR THE INTERNET OF THINGS

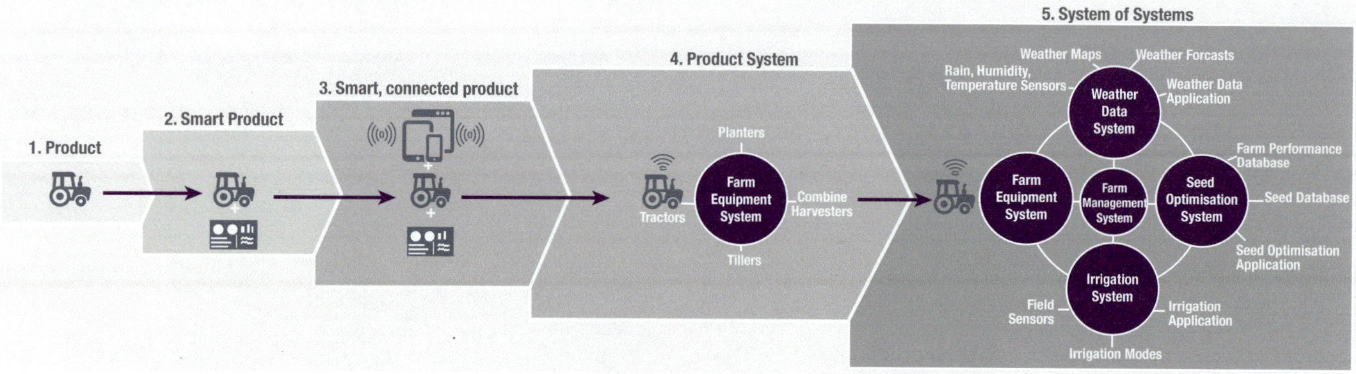

Source: [193]

Total system performance will become more and more important, and companies that address this market need are in better positions to drive this process and gain value [193].

The Transport Network Strategic Investment Tool (TRANSIT), developed by the CSIRO, helps make informed decisions for roads in Northern Australia. TRANSIT helps reduce travel times, and therefore also the stress on the driver and animals, as well as reducing distance, fuel, and wear and tear [194].

The widely discussed smart cities now offer huge business opportunities with an estimated market value of US$1.5 trillion by 2020. Figure 64 shows the various sectors within the Smart City market, and their respective CAGR.

FIGURE 64: SMART CITIES BUSINESS OPPORTUNITIES BY MARKET SEGMENTS

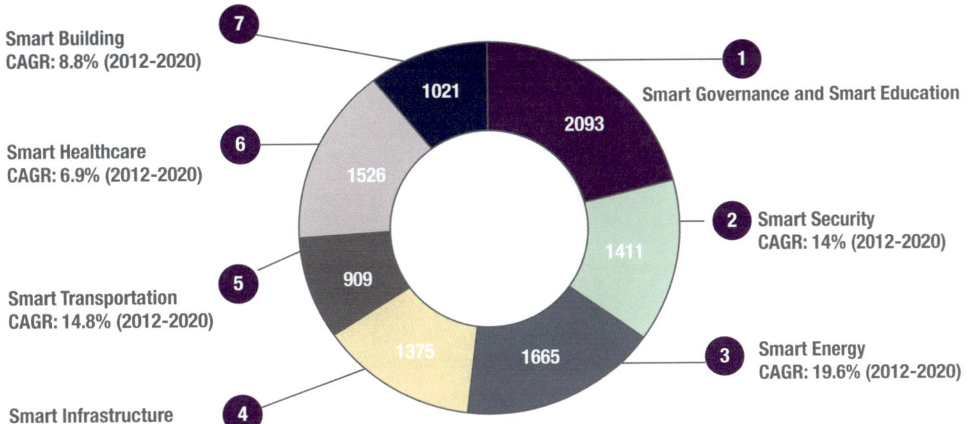

Source: [2]

FIGURE 65: 'THE EDGE' SMART BUILDING, AMSTERDAM

Source: [195]

In Amsterdam, the Edge has implemented smart and green building technology [195]. This 14-storey building has over 6000 LEDs that are used not only for illumination, but that can also transmit information to smartphones at a frequency greater than 2000 flickers per second (unnoticeable to the human eye, but decodable by smartphone cameras – on and off representing binary 0s and 1s). The LEDs, together with motion, temperature and light-level sensors, provide an indoor positioning and monitoring system, and workers can use a smartphone APP (MapIQ) for indoor navigation of the sprawling office building. Technically, real-time information could be provided to workers about their colleagues' location; however, this is not currently offered, as it is considered a privacy minefield. The building has external solar cells and is internally powered, not by conventional 240V electricity cables, but using Ethernet cables. Everything, from coffee machines to hand-towel dispensers, is connected to the internet for analytical purposes. Property managers can analyse the data on how the building is used, and where efficiencies can be gained. The Edge has been called 'a computer with a roof" (see Figure 65) [195].

Figure 66 and Figure 67 show real-life examples of other smart systems in various cities, such as smart transport, smart energy and smart water/waste and social.

FIGURE 66 SMART EMERGING TECHNOLOGIES

Smart Transport

- **Intelligent transport systems**
  Hong Kong: Large-scale traffic management system
- **Congestion charging**
  Singapore: Time-based urban road pricing
- **Smart parking**
  San Francisco: Sensors report free parking spaces and prices are adjusted based on demand
- **Public transport information-sharing**
  Singapore: Inter-modal journey planners and real-time information systems

Smart Energy

- **Smart electricity meters**
  California: Roll out of smart meters since 2006 with > 90% of households accepting
- **Electric vehicle infrastructure**
  France: Staged introduction of obligatory charging stations in new-builds and offices
- **Demand-Response**
  Ontario: "Peaksaver" program to switch off home devices during peak hours
- **Distributed generation integration**
  Lower Saxony: Piloting of dynamic distribution transformers

Smart Water, Waste and Social

- **Smart water meters**
  Toronto: Large roll-out of smart water meters improving metering accuracy and billing efficiency
- **Distribution Network control**
  Sao Palo: Digital water management, real-time system control, automatic leaks warning
- **Remote social infrastructure**
  Boston: Telepresence robots, observing out-patient recovery after discharge from hospital
- **City safety & security**
  Rio de Janerio: Real-time city monitoring and control of municipal service

Source: [196]

FIGURE 67: AUTOMATION IN INFRASTRUCTURE

Automation in operations processes	Driverless trains, as in Dubai, the world´s longest driverless metro network, with 76km in total and 47 stations. Automated border control scheme at Paris-Orly, utilizing biometric fingerprint checks and taking only seconds, thereby saving passengers time and reducing staff. Advanced electricity meters in Nashville, allowing remote and accurate reading, and enabling staff reductions of up to 63%. Unmanned stacking cranes at Abu Dhabi Khalifa Port – 30 cranes fitted with lasers and partially remotely operated, to optimize detection and positioning of containers. Smart street-lighting in four Dutch municipalities, switching off or dimming when no one is walking in the street.
Automation in asset-condition monitoring	Automated ultrasonic testing system on Seto-Ohashi Bridge, with a total length of 13km, to detect welding defects. Wireless monitoring on Jindo Bridge, with 663 wireless sensors constantly analysing vibration, wind and humidity, and reporting anomalies to the central computer. Drones at Autostrade, for inspection of assets and surveillance of highways, particularly in difficult-to-access locations such as bridges in mountainous areas. GE's Cage system, active at 130 sites, measuring sound and electrical signals of sub-sea equipment as early warning for leaks and defect isolation, with 10,000x higher accuracy.
Automation in systems command and control	Rio Operations Centre, which monitors and integrates data from 30 municipal and state agencies and utilities 24/7 under one roof. It monitors everything from weather, traffic, electricity and water to disease outbreaks and emergencies; e.g. enabling the local waste firm to better sequence rubbish collection and reduce fuel use, and enabling the transit agency to redirect traffic promptly to alternative routes to avoid congestion. GE's remote monitoring system of its Jenbacher gas engines. More than 3,400 engines around the world send operating data every 30 seconds to the system, which monitors, diagnoses and predicts their performance. As a result, more than 50% of issues are solved remotely without a service engineer on-site, reducing downtime (which would cost users up to US$ 16,000 a day)

Source: [196]

3 PEOPLE

3.1 HEALTH

In the USA alone, approximately 3000 people die each year from food-borne diseases, and one in six people get sick. Nearly 50% of global food manufacturers have no visibility past the direct suppliers in their supply chain. In Canada the government of Manitoba has implemented a mandatory identification program requiring a unique ID number for any premises where livestock or poultry is grown, kept or disposed of. The system managing this data can track how a disease is spread, contained and eradicated in a much more efficient manner than before [197].

Advances have been reported with the first bionic eye from Second Sight being fitted to a British pensioner suffering from one of the most common forms of blindness, age-related macular degeneration (AMD). This system beams video images directly to the brain, bypassing damaged photoreceptors in his eyes [198].

With sophisticated progress in brain-machine interfaces, a major change in how we operate as humans and as a society is on the horizon. Developments are fuelled by the advanced understanding of the brain (see Figure 68), the use of transcranial-magnetic stimulations or optogenetically enhanced cells – designer cells that can be 'fired'/switched on and off by blue LED lights [199] [200].

FIGURE 68 PATTERNS OF BRAIN ACTIVITY FOR PERCEIVED SELF-LOCATION

Source: [201]

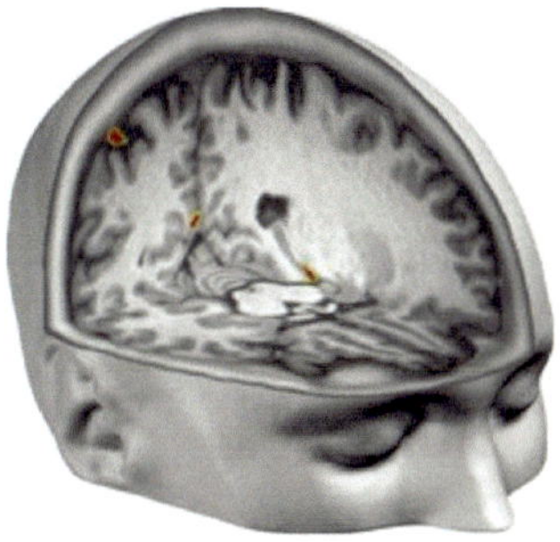

Researchers have reported successful conscious communication of two humans using the internet, enabling telepathy via computer to brain interface. One researcher in France successfully sent words to a researcher in India [202].

It is anticipated that machines will communicate directly and fluently with brains in the not too distant future, changing the way we work. Police and soldiers, or criminals of the 'dark' side can communicate silently and covertly during operations [203]. Planes can be piloted, and missiles shot by mere thoughts [204]. This was demonstrated (in a flight simulator) by TU Munich and TU Berlin. At Arizona State University swarms of small RPAS could be directed by one controller (drone pilot) through focused thoughts [205]. The app MindRDR, Google Glass, together with a Neurosky MindWave sensor (EEG monitor) allows people to take photos with pure thought (brain signals) and to post them on Twitter and Facebook [206].

Braingate (produced by Brown University and Utah company Blackrock Microsystems) is an implant to connect the brains of paralysed people wirelessly with computers with the potential to move robotic arms and wheelchairs [207]. The possibilities of applications for brain-machine interfaces are vast, and will intertwine with spatial technologies.

These advances and new capabilities related to brain science demand new ethical considerations and legislative responses. Two reports related to bioethical considerations for neuroscientific research were commissioned by the US President as part of the BRAIN initiative ("BRAIN Initiative" and "Bioethics Commission Releases Final Neuroscience Report as Part of BRAIN Initiative") [208] [209]. However, in most countries, there is still no legislative framework regulating informed consent or protection of personal information regarding brain-machine interfaces or brain-to-brain interfaces [203]. The relative ease of engineering such non-invasive hardware, and the existing enthusiasm in DIY groups for cognitive enhancement makes it all the more pressing for ethicists, scientists and legislative bodies to work together and prepare society for the upcoming possibilities and challenges, and ensure that the technology is developed with the highest ethical standards [203] [210].

3.2 CROWD SOURCING

Crowdsourced information from the public – very often with spatial components – has become an important means of obtaining data, particularly for crisis management. The earthquake in Nepal in April 2015 again demonstrated the usefulness of geospatial tools for emergency management. A global network of volunteer geeks – digital humanitarians – assisted with vital information to help emergency workers on the ground. Information was gained from the analysis of Twitter tweets with embedded location information or accumulated by active crowd volunteering on platforms such as Tomnod, and the Humanitarian Open Street Map Team (HOT) [211]. Mapbox helped with information on road conditions, and building damage analysis. Shakeslide (from the University of Cardiff) mapped landslide risks. ESRI offered smart maps to visualise impact maps [211]; Facebook and Google helped with locating loved ones in the crisis region [212].

Tools for remote mapping efforts have increased in sophistication since the earthquake in Haiti in 2010. In 2013 crisis mappers assisted with Typhoon Haiyan, and subsequently with Syrian refugees, and the Ebola crisis in West Africa. Since Cyclone Pam (Vanuatu, March 2015) drones were added to the mapping toolkit box. Oblique imagery from RPAS, taken at an angle, can give better assessment results of damage than the traditional nadir satellite imagery [213].

3.3 COLLABORATION

The Deloitte Collaborative Economy report states that collaboration is a key ingredient to success. The report valued collaboration at AUD$ 9.3 billion per year. It found that companies that collaborate are up to five times more likely to grow, twice as likely to be profitable, and twice as likely to outgrow competitors. Yet it also found that half of the businesses surveyed have no collaboration strategy (that their people know of) [214].

3.4 DIGITAL DISRUPTION

Digital disruption will affect the kinds of jobs that will be available in the future. A study by CEDA (Committee for Economic Development of Australia) analysed the likelihood that computerisation will disrupt or replace jobs – see Figure 69.

FIGURE 69: DISTRIBUTION OF JOB CATEGORIES AGAINST PROBABILITY OF COMPUTERISATION

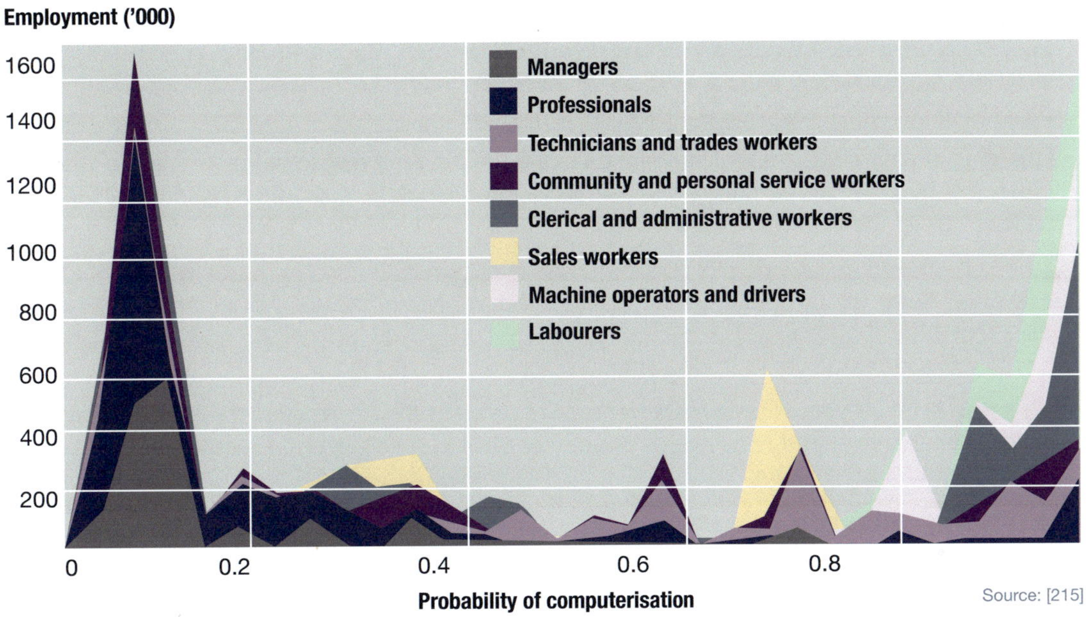

Source: [215]

4 ISSUES

4.1 ARTIFICIAL INTELLIGENCE

Director of Engineering at Google and renown futurist Ray Kurzweil estimates that by 2020 smartphones might have the same storage and processing capacity as a human mind, and that by 2029 'brain software' will be completely functional [66].

Leaders in the IT industry are working on 'cognitive computing', historically known as 'artificial intelligence'. A former IBM Watson project lead, Manoj Saxena, remarked:

> Cognitive computing is the most significant disruption in the evolution of computing since the advent of the Internet. It helps extract patterns and insights from data sources that are almost totally opaque today, what is sometimes known as 'dark data'. Examples include extracting disease insights from healthcare records and social feeds, or finding financial fraud or opportunities from discrete geopolitical, social, and business events. [66]

However, concern that artificial intelligence could become a threat to humans at some time in the future has been voiced by people like Bill Gates, Elon Musk, and Stephen Hawking [216]. During an interview in October 2014, Elon Musk said:

> I think we should be very careful about artificial intelligence. If I had to guess at what our biggest existential threat is, it's probably that. So we need to be very careful. I'm increasingly inclined to think that there should be some regulatory oversight, maybe at the national and international level, just to make sure that we don't do something very foolish. [217]

An article [218] co-written by Tom Dietterich, President of the Association for the Advancement of Artificial Intelligence, and Eric Horvitz, Director of Microsoft Research Lab, outlines three key risks around artificial intelligence:

- Program errors in AI software
- Cyber-attacks on AI systems by criminals, terrorists or government-backed hackers
- Sorcerer's Apprentice Scenarios: AI systems respond to human instructions in unexpected (and possibly dangerous) ways

An open letter of concern has been signed by many well-known personalities in the field [219] asking for focus on research priorities to ensure AI systems are robust and beneficial [220].

4.2 SECURITY AND PRIVACY

According to the report "Next Generation Biometric Market by Technology[14], Function, Application[15] & Geography – Forecasts & Analysis 2014–2020" (published by MarketsandMarkets) the biometric market has a CAGR of 17.6% and will be worth US$23.54 billion by 2020 [221]. Passwords will become a thing of the past. Toronto-based company Nymi, for example, uses the human heartbeat (which the company claims to be unique to each individual) for login [222].

Ten million identities are digitally pickpocketed each year [223]. But security can be enhanced when behavioural biometrics are used to identify users, such as when facilitating a financial transaction on a bank's website. The company Biocatch analyses a person's online behaviour [224]; for instance, when the cursor disappears and the user reacts, some people scan the corners

[14] Technology: Fingerprint, Palm, Face, Iris, Vein, Voice and Signature.
[15] Application: Government, Defence, Travel and Immigration, Home Security, Banking, Consumer Electronics and so on.

of the screen; others wiggle the mouse in circles to retrieve the cursor.

It is easy to use a stolen user ID and password for access, but it is much more difficult to misrepresent behavioural biometrics [225]. Interestingly, a study found that privacy and trust were not the key reasons for biometric login to be adopted by users, but rather usability and social compatibility. Users preferred the Apple fingerprint scan over the constant taking of selfies with the Android smartphone face recognition algorithms [226].

Location data can be used as an extra layer of security. Location data of bank clients' transactions are utilised to combat fraud [227]. Romanian-based company CoSosys has added geofencing to mobile-device management software that allows their customers from hi-tech industries to ensure that visitors do not take pictures in secured areas [228]. Big companies like the Bank of America have run optimisation experiments with high-tech name tags (from Sociometric Solutions) that record the physical location of employees and track how much the person walks, stands, sits, talks and listens [229].

Recently, the US Supreme Court ruled that forcing someone to wear an ankle bracelet for location monitoring qualifies as a Fourth Amendment 'search' [230]. Nevertheless, and perhaps fuelled by Edward Snowden's disclosures, there is a perceived impression by the general public that they are being continuously surveilled by various governments or big-name companies [231] [232] [35].

During a panel discussion at CES 2014 in Las Vegas, Jim Farley, VP of Marketing and Sales at Ford, said: "We know everyone who breaks the law, we know when you're doing it; we have GPS in your car, so we know what you're doing." He then added: "By the way, we don't supply that data to anyone." He later qualified his statement by claiming it was purely hypothetical [233].

Pew Research reported that about 35% of adult social-media users have location information turned off to maintain their privacy on their posts. However, opting out of these settings is not always easy or straightforward [234]. Android apps, according to a recent study by Carnegie Mellon University, track and share the user's location information frequently [230]. Applications like Marauder's Map show where Facebook messages have been sent from, and over time enable the creation of the user's typical location pattern [235].

Cyber crime also exploits the potential for unintended location disclosure. Tech giants and civil liberty groups are concerned over discussions to mandate backdoor access to encrypted communication [236]. Even though such backdoor use could benefit government and law-enforcement agencies, it also creates vulnerability for exploitation.

A 2014 study by the US President's National Security Telecommunications Advisory Committee said there is a 3–5 year window of opportunity to maximise security and minimise risk in adopting the Internet of Everything; otherwise the nation could find itself dealing with adverse consequences for generations [237].

It is possible to hack wireless devices of all kinds, including medical equipment like insulin pumps and pacemakers. Recently, hackers, sitting on a sofa several kilometres away, took over the control of a moving car and crashed it into a ditch [238]. Likewise, a hacker, via the in-flight entertainment system, overwrote code on a large commercial plane's thrust management computer, made the plane climb, and briefly changed course [239]. The hacker, Chris Roberts, a security researcher with One World Labs (now defunct), reported his 'experiment' at a conference and was subsequently interviewed by the FBI [240].

The conflict between security needs and privacy rights is likely to intensify over the next few years as citizens become more aware of what technology and analytics can do. Data about individuals are being collected and meshed together at unprecedented rates. As the Internet of Things advances, connected devices, often invisible, will collect massive amounts of personal data.

In his TED talk on wearable technology Ben Moir (founding member of We:eX), said that no data is truly secure, and the sort of data systems can collect is "terrifying" [241]. The average user has no idea which organisation collects what data or how the data are used, as it is typically too confusing and time-consuming to view this information. Data are often hosted internationally, and privacy laws vary vastly according to each country's jurisdiction.

4.3 GOVERNMENTS' DIGITAL TRANSFORMATION AGENDA

Eggers and Bellman 's recent, "The Journey to Government's digital transformation", notes that governments from Toronto to Seoul are in the midst of a historic transformation as they abandon analogue-operating models in favour of digital systems [242]. Australia's National Innovation and Science Agenda, launched by Prime Minister Malcolm Turnbull, strongly endorses the need for the Digital Transformation of the Australian economy.

Similarly, Digital India, DigitalGov (from USA) and Digital 5, or D5, is a network of leading digital governments with the goal of strengthening the digital economy. The D5 members are New Zealand, South Korea, Israel, the United Kingdom and Estonia. The D5 members are bonded by the principle of openness, and are focused on changing government's relationship with technology by adopting open standards and open-source software, as well as making digital government more effective [243].

Spatial enablement and Location Intelligence together form one of the key components of the digital transformation of the world.

CONCLUSION

There is a sense from this review that the world has a nascent realisation that the digital transformation is providing us with a positioning and location capability that is precise, 'always on', and is tracked, stored and retrievable for instant or future use. Once privacy concerns have been addressed and effective safeguards against the potential dangers are in place, the transformative applications of this capability will be enormous.

REFERENCES

[1] PWC, “Five global megatrends”. [Online]. Available: http://www.pwc.com/gx/en/issues/megatrends/index.jhtml. [Accessed: 22-Apr-2015].

[2] “World’s Top Global Mega Trends to 2025 and Implications to Business, Society and Cultures”, 2014, pp. 1–30.

[3] J. Manyika, M. Chui, J. Bughin, R. Dobbs, P. Bisson and A. Marrs, “Disruptive technologies: Advances that will transform life, business, and the global economy”, pp. 1–176, May 2013.

[4] A. Hellemans, “Jeremy Rifkin on the Internet of Things and the Next Industrial Revolution”, 14-Apr-2015. [Online]. Available: http://spectrum.ieee.org/tech-talk/telecom/internet/jeremy-rifkin-on-the-internet-of-things-and-the-next-industrial-revolution. [Accessed: 29-Apr-2015].

[5] R. Dreischmeier, K. Close and P. Trichet, “The Digital Imperative,” bcgperspectives, 02-Mar-2015. [Online]. Available: https://www.bcgperspectives.com/content/articles/digital_economy_technology_strategy_digital_imperative/. [Accessed: 29-Apr-2015].

[6] S. Perez, “Pokémon Go passed 100 million installs over the weekend”. [Online]. Available: https://techcrunch.com/2016/08/01/pokemon-go-passed-100-million-installs-over-the-weekend/. [Accessed: 11-Aug-2016].

[7] S.E. Needleman, “‘Pokémon Go’ May Leave Rivals Hunting for Gamers’ Attention”. [Online]. Available: http://www.wsj.com/articles/pokemon-go-may-leave-competitors-hunting-for-gamers-attention-1469532601. [Accessed: 29-Jul-2016].

[8] UN, “Transforming our world: the 2030 Agenda for Sustainable Development: Sustainable Development Know”, sustainabledevelopment.un.org, 14-Oct-2015. [Online]. Available: https://sustainabledevelopment.un.org/post2015/transformingourworld. [Accessed: 14-Oct-2015].

[9] Cpauka, “Spatial is crucial to UN Sustainable Development Goals - Spatial Source”, Oct. 2015.

[10] P. Woodgate, I. Coppa and N. Hart, “Global Outlook 2014: Spatial Information Industry”, 05-Nov-2014. [Online]. Available: http://www.crcsi.com.au/assets/45/crcsi-globaloutlook-final-8.pdf. [Accessed: 28-Sep-2015].

[11] H. LeHong and A. Velosa, “Hype Cycle for the Internet of Things, 2014”. gartner.com, 21-Jul-2014. [Online]. Available: http://www.gartner.com/technology/reprints.do?id=1-27LJLAK&ct=150119&st=sb. [Accessed: 10-Jun-2015].

[12] D. Helbing, "The Digital Society: A Summary," SSRN Journal, pp. 1–9, 2014.

[13] D.O. Vermesan and D.P. Friess, "Internet of Things: Converging Technologies for Smart Environments and Integrated Ecosystems", pp. 1–363, Jun. 2013.

[14] AGT, "AGT International Announces IoTA, the First Analytics Platform for the Internet of Things, To Accelerate Development of the IoT". [Online]. Available: http://www.prnewswire.com/news-releases/agt-international-announces-iota-the-first-analytics-platform-for-the-internet-of-things-to-accelerate-development-of-the-iot-279330652.html. [Accessed: 27-May-2015].

[15] B. Wasik, "In the Programmable World, All Our Objects Will Act as One". [Online]. Available: http://www.wired.com/2013/05/internet-of-things-2/. [Accessed: 10-Jun-2015].

[16] A. Tilley, "Samsung Acquires SmartThings, A Fast-Growing Home Automation Startup", http://www.forbes.com/sites/aarontilley/2014/08/14/samsung-smartthings-acquisition-2/#4eafaf917965, 14-Aug-2014. [Online]. Available: http://www.forbes.com/sites/aarontilley/2014/08/14/samsung-smartthings-acquisition-2/. [Accessed: 03-Aug-2015].

[17] Vision Mobile, "IOT MEGATRENDS 2015", pp. 1–24, Feb. 2015.

[18] Myriota, "Myriota | Global Reach for the Internet of Things". [Online]. Available: http://myriota.com. [Accessed: 29-Jul-2015].

[19] Foreshew, Jennifer. "Myriota Aims to Commercialise UniSA M2M Technology." [Online]. Available: http://www.theaustralian.com.au/business/technology/myriota-aims-to-commercialise-unisa-m2m-technology/news-story/9b1c4e0a3952cdc08b25dc5044fdd8e9. [Accessed: 12-Dec-2015].

[20] L. Hickin, "Cut through: How the Internet of Things is sharpening Australia's competitive edge", pp. 1–16, Feb. 2015.

[21] Vandrico, "Wearable Tech Insights", vandrico.com. [Online]. Available: http://vandrico.com/wearables/insights. [Accessed: 17-Jun-2015].

[22] Vandrico, "Wearable Technology Database | Vandrico Inc-2".

[23] Beecham Research, "Wearable Technology Application Chart : Beecham Research", beechamresearch.com. [Online]. Available: http://www.beechamresearch.com/article.aspx?id=20. [Accessed: 24-Jun-2015].

[24] S. Mlot, "Wrist-Based Drone Can Take Off, Snap Selfies", 30-Sep-2014. [Online]. Available: http://au.pcmag.com/consumer-electronics-reviews-ratings/24785/news/wrist-based-drone-can-take-off-snap-selfies. [Accessed: 22-Apr-2015].

[25] J. Anderson and L. Rainie, "The Internet of Things Will Thrive by 2025", 14-May-2014. [Online]. Available: http://www.pewinternet.org/2014/05/14/internet-of-things/. [Accessed: 22-Apr-2015].

[26] E. Strickland, "Medtronic Wants to Implant Sensors in Everyone", 14-Jun-2014. [Online]. Available: http://spectrum.ieee.org/tech-talk/biomedical/devices/medtronic-wants-to-implant-sensors-in-everyone. [Accessed: 22-Apr-2015].

[27] M. Wilson, "From The Designers Of Fitbit, A Digital Tattoo Implanted Under Your Skin", http://www.fastcodesign.com/3036175/from-the-designers-of-fitbit-a-digital-tattoo-implanted-under-your-skin. [Online]. Available: http://www.fastcodesign.com/3036175/from-the-designers-of-fitbit-a-digital-tattoo-implanted-under-your-skin. [Accessed: 11-Dec-2015].

[28] Google ATAP, "Project Jacquard", google.com. [Online]. Available: https://atap.google.com/jacquard. [Accessed: 23-Sep-2015].

[29] AFP, "Google working with Levis on smart clothes", Canberra Times, 30-May-2015.

[30] R. Duus and M. Cooray, "How wearable fitness trackers can take control of your life, for the worse", Canberra Times, 23-Jun-2015.

[31] E. Mandel, "How did the Napa earthquake affect sleep?" 25-Aug-2014. [Online]. Available: https://jawbone.com/blog/napa-earthquake-effect-on-sleep/. [Accessed: 16-Sep-2015].

[32] Infiniti Research, "Global Smart Glasses Market for Augmented Reality 2015–2019 – market research report", reportlinker.com. [Online]. Available: http://www.reportlinker.com/p02911134-summary/Global-Smart-Glasses-Market-for-Augmented-Reality.html. [Accessed: 10-Jun-2015].

[33] P. Lamkin, "The best smartglasses: The Google Glass alternatives", wareable.com. [Online]. Available: http://www.wareable.com/headgear/the-best-smartglasses-google-glass-and-the-rest. [Accessed: 29-Apr-2015].

[34] O. Inbar, "Smart Glasses Market report 2015: towards 1 billion shipments", slideshare.net. [Online]. Available: http://www.slideshare.net/comogard/smart-glasses-2015-for-distribution. [Accessed: 10-Jun-2015].

[35] P. Smith, "Google Glass may never launch in Australia", http://www.afr.com/technology/google-glass-may-never-launch-in-australia-20150118-ka35u, 03-Jun-2015. [Online]. Available: http://www.brw.com.au/p/tech-gadgets/google_glass_may_never_launch_in_7TAedJic03qyZeAPNaliVI. [Accessed: 20-Jan-2015].

[36] I. Lunden, "Augmedix nabs $17M to 'rehumanize' doctor:patient relations using Google Glass | TechCrunch". [Online]. Available: https://techcrunch.com/2016/04/25/augmedix-nabs-17m-to-rehumanize-doctorpatient-relations-using-google-glass/. [Accessed: 27-Jul-2016].

[37] S. Dorfman, "Google Glass Transitions from Consumer Curiosity to Professional Tool", http://www.govtech.com/products/Google-Glass-Transitions-from-Consumer-Curiosity-to-Professional-Tool.html. [Online]. Available: http://www.govtech.com/products/Google-

Glass-Transitions-from-Consumer-Curiosity-to-Professional-Tool.html. [Accessed: 06-May-2015].

[38] L. Nanek, “NeatoCode Techniques, First Facial Recognition Hack for Google Glass”. [Online]. Available: http://neatocode.tumblr.com/post/50292349091/face-rec-on-glass. [Accessed: 03-Jun-2015].

[39] R. Bilton, “New Google Glass hack gives doctors the power of face recognition”. [Online]. Available: http://venturebeat.com/2013/05/13/google-glass-medrefglass-facial-recognition/. [Accessed: 03-Jun-2015].

[40] TED, “Alex Kipman: A futuristic vision of the age of holograms | TED Talk | TED.com”.

[41] Microsoft, “Microsoft HoloLens | Official Site”. [Online]. Available: https://www.microsoft.com/microsoft-hololens/en-us. [Accessed: 24-Aug-2016].

[42] C. Martin, “Microsoft HoloLens UK release date, price, specs & games rumours Smart Glasses Market report 2015: towards 1 billion shipments – New Product – PC Advisor”. [Online]. Available: http://www.pcadvisor.co.uk/new-product/wearable-tech/microsoft-hololens-release-date-price-specs-rumours-uk-development-edition-march-2016-build-friends-onsale-3616014/. [Accessed: 24-Aug-2016].

[43] JPL, “NASA, Microsoft Collaboration Will Allow Scientists to ‘Work on Mars’”. [Online]. Available: http://www.jpl.nasa.gov/news/news.php?feature=4451. [Accessed: 14-Apr-2016].

[44] Microsoft, “Microsoft HoloLens: Partner Spotlight with Trimble – YouTube”. [Online]. Available: Microsoft HoloLens: Partner Spotlight with Trimble – YouTube. [Accessed: 14-Apr-2016].

[45] “Microsoft HoloLens | Official Site”, microsoft.com, 04-Aug-2015. [Online]. Available: https://www.microsoft.com/microsoft-hololens/en-us. [Accessed: 04-Aug-2015].

[46] S. Anthony, “Google invents smart contact lens with built-in camera: Superhuman Terminator-like vision here we come | ExtremeTech”, 15-Apr-2014. [Online]. Available: http://www.extremetech.com/extreme/180571-google-invents-smart-contact-lens-with-built-in-camera-superhuman-terminator-like-vision-here-we-come. [Accessed: 11-Jun-2015].

[47] M. Duffy, “Google's Prototype 'Smart Contact Lens': Measuring Blood Glucose Levels for People with Diabetes”. [Online]. Available: http://www.visionaware.org/blog/visionaware-blog/googles-prototype-smart-contact-lens-measuring-blood-glucose-levels-for-people-with-diabetes-1418/12. [Accessed: 13-Apr-2016].

[48] Science Alert, “Samsung just patented smart contact lenses with a built-in camera,” 08-Apr-2016. [Online]. Available: https://www.sciencealert.com/samsung-just-patented-smart-contact-lenses-with-a-built-in-camera. [Accessed: 13-Apr-2016].

[49] D. Eirdosh, “A Contact Lens for Cows?.” [Online]. Available: http://ieet.org/index.php/IEET/print/6454. [Accessed: 18-Dec-2015].

[50] Discovery Channel, “iOptik on the Discovery Channel 2014 – YouTube”. [Online]. Available: https://www.youtube.com/watch?v=feDE_TMCwVM. [Accessed: 14-Apr-2016].

[51] E. Ackerman, “Innovega Delivers the Wearable Displays that Science Fiction Promised”, http://spectrum.ieee.org/tech-talk/consumer-electronics/audiovideo/innovega-delivers-the-wearable-displays-that-science-fiction-promised, 09-Jan-2014. [Online]. Available: http://spectrum.ieee.org/tech-talk/consumer-electronics/audiovideo/innovega-delivers-the-wearable-displays-that-science-fiction-promised. [Accessed: 11-Jun-2015].

[52] Sawh, Michael, “Daqri Smart Helmet Gives You X-Ray-Like Vision to See Inside Objects.” [Online]. Available: http://www.wareable.com/wearable-tech/daqri-smart-helmet-gives-you-x-ray-like-vision-to-see-inside-objects-2140. [Accessed: 31-Aug-2016].

[53] D. Sung, “50 wearable gamechangers for 2015”, 10-Jun-2015. [Online]. Available: wareable.com. [Accessed: 22-Jun-2016].

[54] Skully, “SKULLY Store: AR-1.” [Online]. Available: https://www.skully.com/store. [Accessed: 09-Mar-2016].

[55] S. Buhr, “AR helmet startup Skully goes down in flames”. [Online]. Available: https://techcrunch.com/2016/07/26/once-promising-ar-motorcycle-helmet-startup-skully-has-crashed-and-burned/. [Accessed: 29-Jul-2016].

[56] Skully, “SKULLY AR-1,” skully.com. [Online]. Available: http://www.skully.com/#features. [Accessed: 05-Aug-2015].

[57] Reuters, “Sony undercuts Oculus with US$399 PlayStation VR”. [Online]. Available: http://www.crn.com.au/news/sony-undercuts-oculus-with-us399-playstation-vr-417031. [Accessed: 17-Mar-2016].

[58] T. Danova, “Virtual reality is the next big consumer tech platform”, linkedin.com. [Online]. Available: https://www.linkedin.com/pulse/virtual-reality-next-big-consumer-tech-platform-tony-danova. [Accessed: 17-Jun-2015].

[59] J. Bezerra, W. Bock and F. Candelon, “The Mobile Revolution: How Mobile Technologies Drive a Trillion-Dollar Impact”. [Online]. Available: https://www.bcgperspectives.com/content/articles/telecommunications_technology_business_transformation_mobile_revolution/. [Accessed: 29-Apr-2015].

[60] FLIR, “FLIR ONE™ Personal Thermal Imager by FLIR® – See the Heat”, flir.com, 20-Aug-2014. [Online]. Available: http://www.flir.com/flirone/explore.cfm. [Accessed: 22-Apr-2015].

[61] Einsteinworld, “einstein™Tablet Technology for K-12 STEM Education Tool”, einsteinworld.com. [Online]. Available: http://einsteinworld.com/product/einstein-tablet/. [Accessed: 05-Aug-2015].

[62] Simband, “Samsung Simband”. [Online]. Available: https://www.simband.io. [Accessed: 20-Jan-2016].

[63] Google ATAP, “Project Tango – Google,” google.com. [Online]. Available: https://www.google.com/atap/project-tango/about-project-tango/. [Accessed: 21-Jul-2015].

[64] Lenovo, “Lenovo and Google Partner on New Project Tango Device | Lenovo Newsroom”, 13-Apr-2016. [Online]. Available: http://news.lenovo.com/news-releases/lenovo-and-google-partner-on-new-project-tango-device.htm. [Accessed: 24-Aug-2016].

[65] Google ATAP, “Meet Ara, the modular phone”. [Online]. Available: https://atap.google.com/ara/. [Accessed: 24-Aug-2016].

[66] W. Fox, “Technology Trends Blog | 2013 | 2014 | 2015 | 2020 | 2050 | Singularity Blog”, futuretimeline.net. [Online]. Available: http://www.futuretimeline.net/blog/computers-internet-blog.htm#.VUBoT1Yobwl. [Accessed: 29-Apr-2015].

[67] MIT Technology Review, “Why Apple Pay Is a Huge Milestone in Payments”, Jul. 2015.

[68] C. Neagle, “Broken NFC terminals, lack of retail support stifling Apple Pay usage”. http://www.networkworld.com/article/2904098/opensource-subnet/broken-nfc-terminals-lack-of-retail-support-stifling-apple-pay-usage.html, 31-Mar-2015. [Online]. Available: http://www.networkworld.com/article/2904098/opensource-subnet/broken-nfc-terminals-lack-of-retail-support-stifling-apple-pay-usage.html. [Accessed: 19-Aug-2015].

[69] IDC, “IDC: Smartphone OS Market Share 2015, 2014, 2013, and 2012”, idc.com. [Online]. Available: http://www.idc.com/prodserv/smartphone-os-market-share.jsp. [Accessed: 05-Aug-2015].

[70] J. Corwin, “bcg.perspectives – Delivering Digital Infrastructure” bcgperspectives.com, 11-Feb-2015. [Online]. Available: https://www.bcgperspectives.com/content/articles/digital_economy_telecommunications_delivering_digital_infrastructure/?utm_source=2015JanDvs&utm_medium=Email&utm_campaign=Ealert. [Accessed: 11-Jun-2015].

[71] InternetLiveStats, “Number of Internet Users (2016) – Internet Live Stats”. [Online]. Available: http://www.internetlivestats.com/internet-users/. [Accessed: 22-Jun-2016].

[72] Ericsson, “Ericsson Mobility Report June 2015”, pp. 1–28, May 2015.

[73] W. Shanklin, “Iridium Go! hotspot can turn your smartphone into a satellite phone”, 04-Feb-2014. [Online]. Available: http://www.gizmag.com/iridium-go-satellite-hotspot/30711/. [Accessed: 22-Apr-2015].

[74] H. Williams, "NBN Has Launched Its Sky Muster Broadband Service For Regional Areas | Gizmodo Australia", 29-Apr-2016. [Online]. Available: http://www.gizmodo.com.au/2016/04/nbn-has-launched-its-sky-muster-broadband-service-for-regional-areas/. [Accessed: 02-May-2016].

[75] Akamai, "Q1 2016 State of the Internet – Connectivity Report | Akamai", Jun. 2016, pp. 1–64.

[76] ABS, "8153.0 – Internet Activity, Australia, December 2015", Apr. 2016. [Online]. Available: http://www.abs.gov.au/ausstats/abs@.nsf/Latestproducts/8153.0Main%20Features4December%202015?opendocument&tabname=Summary&prodno=8153.0&issue=December%202015&num=&view=. [Accessed:13-Sep-2016].

[77] R. Tucker, "Bringing Australia's Broadband Network Into The 21st Century", events.unimelb.edu.au. [Online]. Available: https://events.unimelb.edu.au/recordings/181-bringing-australia-s-broadband-network-into-the-21st-century. [Accessed: 10-Sep-2015].

[78] N. Czernich, O. Falck, T. Kretschmer and L. Woessmann, "Broadband Infrastructure and Economic Growth", http://onlinelibrary.wiley.com/doi/10.1111/j.1468-0297.2011.02420.x/abstract, 12-May-2011. [Online]. Available: http://onlinelibrary.wiley.com/enhanced/doi/10.1111/j.1468-0297.2011.02420.x/. [Accessed: 10-Sep-2015].

[79] C. Kongaut, I. K. Rohman, and E. Bohlin, "The economic impact of broadband speed: Comparing between higher and lower income countries", pp. 1–32, Sep. 2014.

[80] ITU, "ITC: Facts and Figures 2016", 07-Jul-2016. [Online]. Available: https://www.itu.int/en/ITU-D/Statistics/Documents/facts/ICTFactsFigures2016.pdf. [Accessed: 27-Jul-2016].

[81] S. Curtis, "Avanti satellites fly to connect untapped African market to internet", canberratimes.com.au. [Online]. Available: http://www.canberratimes.com.au/digital-life/digital-life-news/avanti-satellites-fly-to-connect-untapped-african-market-to-internet-20150722-gib0e1. [Accessed: 23-Jul-2015].

[82] P. Dockrill, "Facebook is preparing its Internet-beaming drone for maiden launch", 31-Jul-2015. [Online]. Available: http://www.sciencealert.com/facebook-is-preparing-its-internet-beaming-drone-for-maiden-launch. [Accessed: 26-Aug-2015].

[83] A. Vance, "The New Space Race: One Man's Mission to Build a Galactic Internet", 23-Jan-2015. [Online]. Available: http://www.bloomberg.com/news/features/2015-01-22/the-new-space-race-one-man-s-mission-to-build-a-galactic-internet-i58i2dp6. [Accessed: 27-May-2015].

[84] S. Clark, "OneWeb selects Airbus to build 900 Internet satellites – Spaceflight Now", 15-Jun-2015. [Online]. Available: https://spaceflightnow.com/2015/06/15/oneweb-selects-airbus-to-build-900-internet-satellites/. [Accessed: 13-Apr-2016].

[85] A. Knapp, "Branson-Backed OneWeb Raises $500 Million To Build Satellite Internet – Forbes", forbes.com, 25-Jun-2015.

[86] B. Crew, "Google's Internet balloons will soon connect all of Sri Lanka with Wi-Fi – ScienceAlert", sciencealert.com, 30-Jul-2015. [Online]. Available: http://www.sciencealert.com/google-s-internet-balloons-will-soon-connect-all-of-sri-lanka-with-wi-fi. [Accessed: 26-Aug-2015].

[87] C. O'Brien, "In the race between Google and Facebook over global connectivity, everyone is winning", http://venturebeat.com/2015/03/05/in-the-race-between-google-and-facebook-over-global-connectivity-everyone-is-winning/, 05-Mar-2015. [Online]. Available: http://venturebeat.com/2015/03/05/in-the-race-between-google-and-facebook-over-global-connectivity-everyone-is-winning/. [Accessed: 27-May-2015].

[88] F. Khan, "Mobile Internet from the Heavens", pp. 1–7, Jul. 2015.

[89] B. Crew, "Samsung says we can achieve Earth-wide Internet via 4,600 micro-satellites – ScienceAlert", sciencealert.com, 26-Aug-2015. [Online]. Available: http://www.sciencealert.com/samsung-says-we-can-achieve-earth-wide-internet-through-4-600-micro-satellites. [Accessed: 26-Aug-2015].

[90] Space Foundation, "The Space Report 2015: The Authoritative Guide to Global Space Activity", pp. 1–7, Jul. 2015.

[91] OECD, "The Space Economy at a Glance 2014", pp. 1–146, Nov. 2014.

[92] Acil Allen Consulting, "THE VALUE OF EARTH OBSERVATIONS FROM SPACE TO AUSTRALIA", pp. 1–140, Dec. 2015.

[93] UCS, "UCS Satellite Database | Union of Concerned Scientists".

[94] C. LaFleur, "The Spacecrafts Encyclopedia", claudelafleur.qc.ca. [Online]. Available: http://claudelafleur.qc.ca/Spacecrafts-index.html#Military. [Accessed: 21-Oct-2015].

[95] Planet4589, "Sat Stats". [Online]. Available: http://planet4589.org/space/log/stats1.html. [Accessed: 28-Oct-2015].

[96] UCS, "UCS Satellite Database | Union of Concerned Scientists", ucsusa.org, 13-May-2015. [Online]. Available: http://www.ucsusa.org/nuclear_weapons_and_global_security/solutions/space-weapons/ucs-satellite-database.html#.VVLE7FYobwl. [Accessed: 13-May-2015].

[97] DigitalGlobe, "A Detailed View of the Ground with 30-centimeter Satellite Imagery", 08-Mar-2015. [Online]. Available: http://spacenews.com/a-detailed-view-of-the-ground-with-30-centimeter-satellite-imagery/. [Accessed: 27-May-2015].

[98] A.S. Belward and J.O. Skøien, "Who launched what, when and why; trends in global land-cover observation capacity from civilian earth observation satellites", ISPRS Journal of Photogrammetry and Remote Sensing, vol. 103, no. C, pp. 115–128, May 2015.

[99] Euroconsult, "Earth observation, EO, research." [Online]. Available: http://www.euroconsult-ec.com/shop/earth-observation/74-satellite-based-earth-observation-market-prospects-to-2024.html. [Accessed: 04-May-2016].

[100] The Economist, "Nanosats are go!" 07-Jun-2014. [Online]. Available: http://www.economist.com/news/technology-quarterly/21603240-small-satellites-taking-advantage-smartphones-and-other-consumer-technologies. [Accessed: 11-Jun-2015].

[101] UNSW, "QB50 – An International Network of 50 CubeSats | The Australian Centre for Space Engineering Research". [Online]. Available: http://www.acser.unsw.edu.au/QB50. [Accessed: 23-Jun-2016].

[102] Urthecast, "UrtheCast – See the planet. Open the world". [Online]. Available: https://www.urthecast.com/live. [Accessed: 21-Oct-2015].

[103] S. Clark, "Spaceflight Now | Breaking News | UrtheCast says it will be largest commercial ISS user", spaceflightnow.com, 21-Oct-2014. [Online]. Available: http://spaceflightnow.com/news/n1410/03urthecast/#.Vicehc4obdl. [Accessed: 21-Oct-2015].

[104] A. Wallace, "Airbus and Hexagon partner for high-res imagery", http://www.spatialsource.com.au/airbus-and-hexagon-partner-for-high-res-imagery/. [Online]. Available: http://www.spatialsource.com.au/2015/06/02/airbus-and-hexagon-partner-for-high-res-imagery/. [Accessed: 03-Jun-2015].

[105] Spaceflight Now, "Spaceflight Now | Tracking Station | Launch log", spaceflightnow.com. [Online]. Available: http://spaceflightnow.com/tracking/launchlog.html. [Accessed: 21-Oct-2015].

[106] Spaceflight Now, "Launch Schedule | Spaceflight Now", spaceflightnow.com, 21-Oct-2015. [Online]. Available: https://spaceflightnow.com/launch-schedule/. [Accessed: 21-Oct-2015].

[107] GPS World, "Upcoming GNSS satellite launches: GPS World", 01-Aug-2016. [Online]. Available: http://gpsworld.com/resources/upcoming-gnss-satellite-launches/. [Accessed: 01-Aug-2016].

[108] Whatech, "Global real-time location system market to grow at a CAGR of 37.41% between 2015–2019. – WhaTech", whatech.com, 01-Apr-2015. [Online]. Available: http://www.whatech.com/market-research-reports/press-release/it/45095-global-real-time-location-system-market-to-grow-at-a-cagr-of-37-41-between-2015-2019?lf1=9201240615b941411202188a38054913. [Accessed: 11-Jun-2015].

[109] D.A. Divis, "Study: GPS Contributed More Than $68 Billion to the U.S. Economy | Inside GNSS", insidegnss.com. [Online]. Available: http://www.insidegnss.com/node/4535. [Accessed: 21-Jul-2015].

[110] A. Wallace, “GDA2020 proposal plan under development”, http://www.spatialsource.com.au/gda2020-proposal-plan-under-development/. [Online]. Available: http://www.spatialsource.com.au/2015/05/05/gda2020-proposal-plan-under-development/. [Accessed: 06-May-2015].

[111] S. Chester, “Japan’s QZSS GNSS to benefit Australia – Spatial Source”, spatialsource.com.au, 14-Apr-2015. [Online]. Available: http://www.spatialsource.com.au/2015/04/14/japans-qzss-gnss-to-benefit-australia/. [Accessed: 19-Dec-2015].

[112] D.A. Divis, “PNT Advisory Board Debates Critical Infrastructure Designation for GPS | Inside GNSS”, insidegnss.com. [Online]. Available: http://www.insidegnss.com/node/4536. [Accessed: 21-Jul-2015].

[113] J. Lendino, “DARPA to re-invent GPS navigation without the use of satellites", ExtremeTech. [Online]. Available: http://www.extremetech.com/extreme/202111-darpa-to-re-invent-gps-navigation-without-satellites. [Accessed: 10-Jun-2015].

[114] insidegnss.com. [Online]. Available: http://www.insidegnss.com. [Accessed: 21-Jul-2015].

[115] G. Gibbons, “South Korea Relaunches Its eLoran Program | Inside GNSS”, 12-Jan-2016.

[116] Locata, “Synchronized Ground Networks Usher in Next-Gen GNSS”, pp. 1–8, Oct. 2013.

[117] Rizos C; Kealy A; Li B; Choudhury M; Choy S; Feng Y, 2014, 'Locata's VRay™ antenna technology - Multipath mitigation for indoor positioning', in 27th International Technical Meeting of the Satellite Division of the Institute of Navigation, ION GNSS 2014, pp. 1629 - 1635

[118] Locata, “White Sands Missile Range LocataNet | Locata”, locata.com, 27-Aug-2013. [Online]. Available: http://www.locata.com/article/white-sands-missile-range-locatanet-new/. [Accessed: 19-Aug-2015].

[119] A. Wallace, “Australia’s Locata to direct NASA’s UAV tests.” [Online]. Available: http://www.spatialsource.com.au/australias-locata-to-direct-nasas-uav-tests/. [Accessed: 09-Sep-2015].

[120] A. Wallace, “Locata leads revolution in positioning – Spatial Source”. [Online]. Available: http://www.spatialsource.com.au/2016/08/locata-leads-revolution-non-gps-positioning/. [Accessed: 03-Aug-2016].

[121] R. Mautz, “Indoor Positioning Technologies”, 12-Jun-2012. [Online]. Available: http://e-collection.library.ethz.ch/eserv/eth:5659/eth-5659-01.pdf. [Accessed: 08-Aug-2016].

[122] UT News, “New Centimeter-Accurate GPS System Could Transform Virtual Reality and Mobile Devices”, 06-May-2015. [Online]. Available: http://newsutexasedu/2015/05/05/engineers-develop-centimeter-accurate-gps-system. [Accessed: 17-Jun-2015].

[123] Business Wire, "Research and Markets: Global LBS Market Report 2015-2019 – Increasing Use of LBS to Monitor Consumer Behaviour", 24-Jun-2015. [Online]. Available: http://www.businesswire.com/news/home/20150624005679/en/Research-Markets-Global-LBS-Market-Report-2015-2019. [Accessed: 19-Aug-2015].

[124] Al Urbanski, "Study: Mobile Campaigns Drive Twice the Sales of Desktop – Direct Marketing News", 11-May-2015.

[125] 4Info, "Do Your Mobile Metrics Measure Up?" pp. 1–22, Jun. 2015.

[126] N. Eddy, "Location-Based Advertising Market to Hit Nearly $15 Billion by 2018", 28-Apr-2014. [Online]. Available: http://www.eweek.com/mobile/location-based-advertising-market-to-hit-nearly-15-billion-by-2018.html. [Accessed: 22-Apr-2015].

[127] Technavio, "Global LBS Market 2015–2019 – Research and Markets", researchandmarkets.com, 10-Jun-2015. [Online]. Available: http://www.researchandmarkets.com/research/nwllfw/global_lbs_market. [Accessed: 19-Aug-2015].

[128] ABIResearch, "BLE Beacons", abiresearch.com, 2015. [Online]. Available: https://www.abiresearch.com/pages/ble-beacons/. [Accessed: 19-Aug-2015].

[129] A. Buczkowski, "With iPhone 6 and iOS8 Apple starts to be serious about indoor location", geoawesomeness.com, 17-Sep-2014. [Online]. Available: http://geoawesomeness.com/with-iphone-6-and-ios8-apple-starts-to-be-serious-about-indoor-location/. [Accessed: 22-Apr-2015].

[130] M. Ballve, "Commercial drones have leapt far ahead of regulators and are already entrenched in a handful of massive industries", 28-Feb-2015. [Online]. Available: http://www.businessinsider.com.au/uav-or-commercial-drone-market-forecast-2015-2. [Accessed: 10-Dec-2015].

[131] ABIResearch, "Small Unmanned Aerial Systems Market Exceeds US$8.4 Billion by 2019", abiresearch.com, 05-Jan-2015. [Online]. Available: https://www.abiresearch.com/press/small-unmanned-arial-systems-market-exceeds-us84-b/. [Accessed: 29-Apr-2015].

[132] D. Chow, "The Future of Drones: Sky-High Hopes vs. Regulatory Realities". http://www.livescience.com/45242-future-of-drones.html, 30-Apr-2014.

[133] Skytrails, "THE CIVILIAN DRONE LANDSCAPE – SUMMER 2014", 20-Jun-2014. [Online]. Available: http://dronelife.com/wp-content/uploads/2014/08/230493945-skytrails-presents-the-civilian-drone-landscape-summer-2014.pdf. [Accessed: 24-Jun-2015].

[134] X: Moonshot Factory, "Introducing Project Wing", Sep. 2014.

[135] "Surf Lifesaving UAV wins innovation award – Spatial Source", spatialsource.com.au, 17-Sep-2014. [Online]. Available: http://www.spatialsource.com.au/2014/09/16/surf-lifesaving-uav-wins-innovation-award/. [Accessed: 11-Jun-2015].

[136] ASM, “Crowdsourcing Effort Helps Map the Sundarbans Oil Spill”, 20-Feb-2015. [Online]. Available: https://asmmag.com/features/feature/7425-crowdsourcing-effort-helps-map-the-sundarbans-oil-spill.html. [Accessed: 11-Jun-2015].

[137] M. Tully, “Unmanned Aerial Systems & Remote-Sensing”, [Online]. Available: http://aerialservicesinc.com/wp-content/uploads/2014/07/AerialServices_UASGuide.pdf. [Accessed: 22-Apr-2015].

[138] RPASTrainingSolutions, “Drone Laws in Australia – We explain CASA CASR 101 laws for RPAS in Australia”. [Online]. Available: https://www.rpastraining.com.au/casr-101-uav-drone-legal-or-illegal. [Accessed: 23-Sep-2015].

[139] D. Terdiman, “Google's Project Wing Swoops Into Drone Air Traffic Control Conversation”, 29-Jul-2015. [Online]. Available: http://www.fastcompany.com/3049186/tech-forecast/googles-project-wing-swoops-into-drone-air-traffic-control-conversation. [Accessed: 16-Sep-2015].

[140] K. Atherton, “AirMap Is A Sky Atlas For Drone Users”. [Online]. Available: http://www.popsci.com.au/robots/drones/airmap-is-a-sky-atlas-for-drone-users,403484. [Accessed: 13-May-2015].

[141] M. Corcoran, “Chinese manufacturer programs Phantom drones with no-fly zones to protect Australian airports”, http://www.abc.net.au/news/2014-04-14/chinese-made-drones-programmed-with-no-fly-zones/5388356, 14-Apr-2014. [Online]. Available: http://www.abc.net.au/news/2014-04-14/chinese-made-drones-programmed-with-no-fly-zones/5388356. [Accessed: 22-Apr-2015].

[142] J. Hsu, “DJI Phantom 3 Drone Gets Control Upgrades, Powerful Cameras”, http://spectrum.ieee.org/automaton/robotics/drones/dji-phantom-3-camera-drones, 13-Apr-2015.

[143] CASA, “Part 101 Amendments – Cutting red tape for remotely piloted aircraft”. [Online]. Available: https://www.casa.gov.au/aircraft/standard-page/part-101-amendments-cutting-red-tape-remotely-piloted-aircraft. [Accessed: 01-Aug-2016].

[144] Airservices, “Management of Remotely Piloted Aircraft Systems (RPAS) in ATM operations,” pp. 1–32, Jun. 2016.

[145] R. Clothier and J. Roberts, “New relaxed drone regulations will help the industry take off”. [Online]. Available: http://theconversation.com/new-relaxed-drone-regulations-will-help-the-industry-take-off-57201. [Accessed: 01-Aug-2016].

[146] Airservices, “OneSKY Australia – one step closer”. [Online]. Available: https://newsroom.airservicesaustralia.com/releases/onesky-australia-one-step-closer. [Accessed: 23-Mar-2016].

[147] P. Pullen, “Eyes in the skyInquiry into drones and the regulation of air safety and privacy”, pp. 1–76, Jul. 2014.

[148] J. Hsu, “Aero-X hoverbike set to take off in 2017”, http://spectrum.ieee.org/tech-talk/transportation/alternative-transportation/aerox-hoverbike-goes-on-sale-in-2017, 13-May-2014.

[149] B. Lufkin, “The Military Is Building A Hoverbike”, http://www.gizmodo.com.au/2015/06/the-military-is-building-a-hoverbike/, 23-Jun-2015.

[150] “FAA Grants Terrafugia sUAS Exemption for TF-X™ Flight Testing | Terrafugia”, 17-Dec-2015. [Online]. Available: http://www.terrafugia.com/news/faa-grants-terrafugia-uas-exemption-tf-x-flight-testing. [Accessed: 17-Dec-2015].

[151] B. Popper, “Ehang's autonomous helicopter promises to fly you anywhere, no pilot required”, http://www.theverge.com/2016/1/6/10721654/electric-self-flying-quadcopter-ehang-184-ces-2016, 06-Jan-2016. [Online]. Available: http://www.theverge.com/2016/1/6/10721654/electric-self-flying-quadcopter-ehang-184-ces-2016. [Accessed: 13-Apr-2016].

[152] K. RUSSELL, “An Israeli Company Is Testing Flying Ambulance Drones | Business Insider”, 29-Aug-2016. [Online]. Available: http://www.businessinsider.com.au/israel-testing-flying-ambulance-drones-2014-1. [Accessed: 29-Aug-2016].

[153] Wikipedia, “Open Data Institute – Wikipedia, the free encyclopedia”, en.wikipedia.org. [Online]. Available: https://en.wikipedia.org/wiki/Open_Data_Institute. [Accessed: 26-Aug-2015].

[154] Symbios, “Risks of Data Supply of Earth Observations from Space for Australia”, 09-Dec-2015. [Online]. Available: http://www.crcsi.com.au/assets/Program-2/Risks-of-Data-Supply-of-Earth-Observations-from-Space-for-Australia-Symbios-Communications-Final.pdf. [Accessed: 16-Mar-2016].

[155] ESA, “Missions – Sentinel Online”. [Online]. Available: https://sentinel.esa.int/web/sentinel/missions. [Accessed: 24-Jun-2016].

[156] A. Wallace, “CSIRO signs cooperative agreement with ESA – Spatial Source”.

[157] Department of the Prime Minister and Cabinet, “Public Sector Data Management”, pp. 1–55, Dec. 2015.

[158] Australian Government, “The National Map”. [Online]. Available: https://nationalmap.gov.au. [Accessed: 01-Aug-2016].

[159] Victorian Department of Premier and Cabinet, “Victorian Government Data Directory”. [Online]. Available: https://www.data.vic.gov.au. [Accessed: 01-Aug-2016].

[160] A. Wallace, “VicRoads' open data clears the road to safety – Spatial Source”, spatialsource.com.au. [Online]. Available: http://www.spatialsource.com.au/2015/09/08/vicroads-open-data-clears-the-road-to-safety/. [Accessed: 09-Sep-2015].

[161] M. Goodchild, H. Guo, A. Annoni, L. Bian, K. de Bie, F. Campbell, M. Craglia, M. Ehlers, J. van Genderen, D. Jackson, A. Lewis, M. Pesaresi, G. Remetey-Fulopp, R. Simpson, A. Skidmore, C. Wang, and P. Woodgate, "Next-generation Digital Earth," 14-Apr-2012. [Online]. Available: http://www.crcsi.com.au/assets/Resources/72bbdfad-bdf0-402e-b9f5-67ae8198ac4f.pdf. [Accessed: 10-Jun-2016].

[162] IFE Visualisation and eResearch, "G20 Globe: The Cube Globe on Vimeo", 26-Aug-2015. [Online]. Available: https://vimeo.com/112011647. [Accessed: 26-Aug-2015].

[163] OpenDigitalEarthFoundation, "Queensland G20 Globe", 16-Jun-2015. [Online]. Available: https://digitalearthglobe.org/home/news-2/. [Accessed: 26-Aug-2015].

[164] J. Keysers, "Review of Digital Globes 2015", pp. 1–24, Mar. 2015 CRCSI.

[165] C. Kiehle, K. Greve, and C. Heier, "Requirements for Next Generation Spatial Data Infrastructures-Standardized Web Based Geoprocessing and Web Service Orchestration", Transactions in GIS, vol. 11, no. 6, pp. 819–834, Dec. 2007.

[166] B. Regalia, K. Janowicz, and S. Gao, "VOLT: A Provenance-Producing, Transparent SPARQL Proxy for the On-Demand Computation of Linked Data and its Application to Spatiotemporally Dependent Data", in The Semantic Web. Latest Advances and New Domains, vol. 9678, no. 32, Cham: Springer International Publishing, 2016, pp. 523–538.

[167] K. Janowicz, S. Schade, A. Bröring, C. Keßler, P. Maué, and C. Stasch, "Semantic Enablement for Spatial Data Infrastructures", Transactions in GIS, vol. 14, no. 2, pp. 111–129, Apr. 2010.

[168] W3, "RDF - Semantic Web Standards". [Online]. Available: https://www.w3.org/RDF/. [Accessed: 18-Jun-2016].

[169] OGC, "GeoSPARQL - A Geographic Query Language for RDF Data". [Online]. Available: http://www.opengeospatial.org/standards/geosparql. [Accessed: 18-Jun-2016].

[170] P. Varadharajulu, M. Azeem Saqiq, F. Yu, D. A. McMeekin, G. West, L. Arnold, and S. Moncrieff, "SPATIAL DATA SUPPLY CHAINS", Int. Arch. Photogramm. Remote Sens. Spatial Inf. Sci., vol. 4, pp. 41–45, 2015.

[171] Sensor & Systems, "Sensors & Systems Ten Predictions for 2015", sensorsandsystems.com. [Online]. Available: http://sensorsandsystems.com/sensors-systems-ten-predictions-for-2015/. [Accessed: 23-Sep-2015].

[172] G. Zeiss, "Standards for the convergence of BIM, 3D and geospatial – Geospatial World", 22-Jun-2014. [Online]. Available: http://www.geospatialworld.net/standards-for-the-convergence-of-bim-3d-and-geospatial/. [Accessed: 11-Jun-2015].

[173] DLT Solutions, "Top 7 BIM trends", 24-Mar-2015. [Online]. Available: https://afd34ee8b0806295b5a7-9fbee7de8d51db511b5de86d75069107.ssl.cf1.rackcdn.com/DLT_855-BIMTrendsWhitepaper.pdf. [Accessed: 06-May-2015].

[174] A. Wallace, "Leica Geosystems collaborates with Autodesk – Spatial Source", spatialsource.com.au, 10-Jun-2015. [Online]. Available: http://www.spatialsource.com.au/2015/06/leica-geosystems-collaborates-with-autodesk/. [Accessed: 24-Aug-2016].

[175] N. Mueller, A. Lewis, D. Roberts, S. Ring, R. Melrose, J. Sixsmith, L. Lymburner, A. McIntyre, P. Tan, S. Curnow and A. Ip, "Water observations from space: Mapping surface water from 25 years of Landsat imagery across Australia", Remote Sensing of Environment, vol. 174, no. C, pp. 341–352, Mar. 2016.

[176] A. Sternstein, "CDC Tracks Ebola in West Africa With Cell Phone Location Data", 17-Oct-2014. [Online]. Available: http://mashable.com/2014/10/10/cdc-ebola-cell-phone-tracking/. [Accessed: 22-Apr-2015].

[177] C. Leigh, "CSIRO calls on researchers worldwide to join forces to save honey bees", http://www.csiro.au/en/News/News-releases/2015/Honey-Bee-Health, 25-Aug-2015. [Online]. Available: http://www.csiro.au/en/News/News-releases/2015/Honey-Bee-Health. [Accessed: 23-Sep-2015].

[178] A. Wallace, "Australian researchers predicted Nepal quake", Aug. 2015.

[179] S. Warsmith, "Akron, Ohio, Police to Use New Spatial Video Approach to Supplement Crime Data". [Online]. Available: http://www.govtech.com/dc/articles/Akron-Ohio-Police-to-Use-New-Spatial-video-approach-to-supplement-crime-data.html. [Accessed: 06-May-2015].

[180] e.Republic, "The Candid Camera: Community, Crime-Fighting and the Value of Video", 01-May-2015. [Online]. Available: https://afd34ee8b0806295b5a7-9fbee7de8d51db511b5de86d75069107.ssl.cf1.rackcdn.com/DC15-GT_CASE_STUDY_Motorola_V.PDF. [Accessed: 06-May-2015].

[181] A. Coyne, "Vic police want to shoot GPS trackers at fleeing cars", http://www.itnews.com.au/news/vic-police-want-to-shoot-gps-trackers-at-fleeing-cars-414804, 09-Feb-2016.

[182] C. Matyszczyk, "Police firing GPS tracking 'bullets' at cars during chases – CNET". [Online]. Available: http://www.cnet.com/au/news/police-firing-gps-tracking-bullets-at-cars-during-chases/. [Accessed: 03-Aug-2016].

[183] BBC, "Cows connected to web to boost milk – BBC News", 11-Feb-2015. [Online]. Available: http://www.bbc.com/news/uk-scotland-scotland-business-26705812. [Accessed: 22-Apr-2015].

[184] Moonitor, "Moonitor Cattle Monitoring System", Dec. 2015. http://www.moonitorcows.com [Accessed: 03-Aug-2016].

[185] S. Kumbhar, "MOOnitor selects Globalstar tech chosen for cattle collar to increase beef yields – IoT Now – How to run an IoT enabled business", Oct. 2015.

[186] Cattle Watch, “CattleWatch – Home”, cattle-watch.com. [Online]. Available: http://www.cattle-watch.com/. [Accessed: 21-Dec-2015].

[187] S. Leigo, “Media Release: Managing Australian Cattle Stations From Space”, 21-Oct-2015.

[188] L. Zhao, S. Sun, and X. Wang, “Tracking and Traceability system using livestock iris code in meat supply chain”, ICIC International, pp. 1–12, Jan. 2011.

[189] Surur, “First smartphone with Iris Scanner released”, http://mspoweruser.com/first-smartphone-with-iris-scanner-released/, 03-Aug-2015. [Online]. Available: http://mspoweruser.com/first-smartphone-with-iris-scanner-released/. [Accessed: 17-Dec-2015].

[190] L. Webster, “Australian feeder cattle could be tattooed to indicate stress levels,” 20-Oct-2015. [Online]. Available: http://www.abc.net.au/news/2015-10-20/meat-and-livestock-australia-to-trial-cattle-tattoos/6865550. [Accessed: 17-Dec-2015].

[191] P. Tickle and P. Delaney, “NRM Spatial Information Hub » CRCSI”. [Online]. Available: http://www.crcsi.com.au/research/commissioned-research/nrm-spatial-information-hub/. [Accessed: 23-Mar-2016].

[192] Agersens, “Overview of Fenceless Farming”, 09-Nov-2015. [Online]. Available: http://agersens.com/dairy-and-beef-cattle/. [Accessed: 17-Dec-2015].

[193] M. Porter and J. Heppelmann, “How Smart, Connected Products Are Transforming Competition”, 01-Nov-2014. [Online]. Available: https://hbr.org/2014/11/how-smart-connected-products-are-transforming-competition. [Accessed: 10-Jun-2015].

[194] Getfarming, “Get Farming: Better Roads for Beef Transport in Northern Australia”. [Online]. Available: http://getfarming.com.au/2016/03/09/better-roads-beef-transport-northern-australia/. [Accessed: 24-Aug-2016].

[195] P. Marks, “Office of coded light”, New Scientist. Aug. 2015.

[196] BostonConsultingGroup, “Strategic Infrastructure Steps to Operate and Maintain Infrastructure Efficiently and Effectively”, pp. 1–88, Apr. 2014.

[197] S. Thompson, “From trough to table: mapping the food chain saves lives”, https://www.theguardian.com/sustainable-business/2015/jan/16/trough-table-mapping-food-supply-chain, 16-Jan-2015. [Online]. Available: http://www.theguardian.com/sustainable-business/2015/jan/16/trough-table-mapping-food-supply-chain?CMP=share_btn_tw. [Accessed: 11-Jun-2015].

[198] N. Khomami, “British pensioner first to have central vision restored through bionic eye”, 22-Jul-2015. [Online]. Available: http://www.theguardian.com/uk-news/2015/jul/21/bionic-eye-pensioner-manchester-vision-restored. [Accessed: 16-Sep-2015].

[199] J. Hewitt, "The first thought-activated implant is an exciting hint of things to come | ExtremeTech", 12-Nov-2014. [Online]. Available: http://www.extremetech.com/extreme/193990-the-first-thought-activated-brain-implant-is-an-exciting-hint-of-things-to-come. [Accessed: 06-May-2015].

[200] S. Ramirez, X. Liu, P. A. Lin, J. Suh, M. Pignatelli, R. L. Redondo, T. J. Ryan, and S. Tonegawa, "Creating a False Memory in the Hippocampus", Science, vol. 341, no. 6144, pp. 387–391, Jul. 2013.

[201] J. Dorrier, "Virtual Body Swap Experiment Maps Out-of-Body Illusion in the Brain," 01-May-2015. [Online]. Available: http://singularityhub.com/2015/05/01/virtual-body-swap-experiment-maps-out-of-body-illusion-in-the-brain/. [Accessed: 06-May-2015].

[202] C. Grau, R. Ginhoux, A. Riera, T.L. Nguyen, H. Chauvat, M. Berg, J.L. Amengual, A. Pascual-Leone, and G. Ruffini, "Conscious Brain-to-Brain Communication in Humans Using Non-Invasive Technologies", PLoS ONE, vol. 9, no. 8, p. e105225, Aug. 2014.

[203] J.B. Trimper, P.R. Wolpe, and K.S. Rommelfanger, "When 'I' becomes 'We': ethical implications of emerging brain-to-brain interfacing technologies", ncbi.nlm.nih.gov, 12-Feb-2014. [Online]. Available: http://www.ncbi.nlm.nih.gov/pmc/articles/PMC3921579/. [Accessed: 24-Jun-2015].

[204] J. Hewitt, "How to fly a plane with your brain (but you really shouldn't – not yet, anyway) | ExtremeTech". [Online]. Available: http://www.extremetech.com/extreme/183146-how-to-fly-a-plane-with-your-brain-but-you-really-shouldnt-not-yet-anyway. [Accessed: 06-May-2015].

[205] ASU, "Using Wireless Interface, Operators Control Multiple Drones by Thinking of Various Tasks". [Online]. Available: https://www.sciencedaily.com/releases/2016/07/160713143744.htm. [Accessed: 03-Aug-2016].

[206] Wikipedia, "MindRDR – Wikipedia, the free encyclopedia", en.wikipedia.org.

[207] A. Regalado, "A Brain-Computer Interface That Works Wirelessly", 14-Jan-2015. [Online]. Available: https://www.technologyreview.com/s/534206/a-brain-computer-interface-that-works-wirelessly/. [Accessed: 06-May-2015].

[208] Whitehouse, "BRAIN Initiative". [Online]. Available: https://www.whitehouse.gov/node/226036. [Accessed: 23-Sep-2015].

[209] M.A. Anderson, "Bioethics Commission Releases Final Neuroscience Report as Part of BRAIN Initiative", 26-Mar-2015. [Online]. Available: http://blog.bioethics.gov/2015/03/26/bioethics-commission-releases-final-neuroscience-report-as-part-of-brain-initiative-focuses-on-controversial-topics-that-must-be-addressed-if-neuroscience-is-to-progress-and-be-applied-ethically/. [Accessed: 23-Sep-2015].

[210] F. Cincotti, D. Mattia, F. Pichiorri and F. Schettin, "BNCI Horizon 2020", pp. 1–11, Oct. 2014.

[211] A. Wallace, “International mapping community help Nepal – Spatial Source”, spatialsource.com.au. [Online]. Available: http://www.spatialsource.com.au/2015/05/05/international-mapping-community-help-nepal/. [Accessed: 06-May-2015].

[212] N.P.A.S.B. Shirley S Wang, “Nepal Aid Workers Helped by Drones, Crowdsourcing – WSJ”, wsj.com. [Online]. Available: http://www.wsj.com/article_email/nepal-aid-workers-helped-by-drones-crowdsourcing-1430483540-lMyQjAxMTE1NzAwMzUwNjM1Wj. [Accessed: 06-May-2015].

[213] “How ‘Crisis Mapping’ Is Shaping Disaster Relief in Nepal”, 06-May-2015. [Online]. Available: http://news.nationalgeographic.com/2015/05/150501-nepal-crisis-mapping-disaster-relief-earthquake/?utm_source=Twitter&utm_medium=Social&utm_content=link_tw20150501news-earthquakedisast&utm campaign=Content&sf8933117=1. [Accessed: 06-May-2015].

[214] Deloitte, “The Collaborative Economy”, pp. 1–20, Jul. 2014.

[215] CEDA, “Australia’s future workforce?” pp. 1–258, May 2015.

[216] S. Dredge, “Artificial intelligence will become strong enough to be a concern, says Bill Gates”. [Online]. Available: https://www.theguardian.com/technology/2015/jan/29/artificial-intelligence-strong-concern-bill-gates. [Accessed: 29-Aug-2016].

[217] A. Strange, “We are ‘summoning the demon’ with AI: technologist Elon Musk”, http://www.smh.com.au/digital-life/digital-life-news/we-are-summoning-the-demon-with-ai-technologist-elon-musk-20141026-11c8ql.html, 27-Oct-2014. [Online]. Available: http://www.canberratimes.com.au/digital-life/digital-life-news/we-are-summoning-the-demon-with-ai-technologist-elon-musk-20141026-11c8ql.html. [Accessed: 11-Jun-2015].

[218] M. Tegmark, “Benefits and Risks of Artificial Intelligence – Medium”, medium.com. [Online]. Available: https://medium.com/@tdietterich/benefits-and-risks-of-artificial-intelligence-460d288cccf3. [Accessed: 09-Sep-2015].

[219] M. Tegmark, “The Future of Life Institute”. [Online]. Available: http://futureoflife.org. [Accessed: 03-Jun-2015].

[220] S. Russell, D. Dewey, and M. Tegmark, “Research priorities for robust and beneficial artificial intelligence”, pp. 1–12, Jan. 2015.

[221] PRNewswire, “Next Generation Biometric Market – 2020 | MarketsandMarkets”, marketsandmarkets.com. [Online]. Available: http://www.marketsandmarkets.com/Market-Reports/next-generation-biometric-technologies-market-697.html. [Accessed: 27-May-2015].

[222] Nymi, “Product Overview | Nymi”. [Online]. Available: https://nymi.com. [Accessed: 03-Aug-2016].

[223] Betabrand, “Betabrand & Norton Create the World's First RFID-Blocking Jeans and Blazer – Betabrand Blog”, 10-Dec-2014. [Online]. Available: https://www.betabrand.com/blog/33531/betabrand-norton-creates-worlds-first-rfid-blocking-jeans-blazer. [Accessed: 29-Apr-2015].

[224] Biocatch, “Behavioral Authentication”. [Online]. Available: http://www.biocatch.com/behavioral-authentication?__hstc=151929382.475c5d0cb6c384e06ce80a758d8db11a.1470201528721.1470201528721.1470201528721.1&__hssc=151929382.1.1470201528721&__hsfp=2809335995. [Accessed: 03-Aug-2016].

[225] B. Blum, “Hackers meet their match”, 27-May-2015. [Online]. Available: http://www.israel21c.org/hackers-meet-their-match/. [Accessed: 04-Jun-2015].

[226] A. De Luca, A. Hang, E. von Zezschwitz and H. Hussmann, “I Feel Like I'm Taking Selfies All Day!” New York, USA, 2015, pp. 1411–1414.

[227] D. McGrath, “Geo-location to Combat Fraud | Unisys Blogs | The Official Unisys Blog,” blogs.unisys.com. [Online]. Available: http://blogs.unisys.com/financialindustryinsights/geo-location-to-combat-fraud/. [Accessed: 02-Sep-2015].

[228] M. Korolov, “Geofencing could add security layer for mobile devices”, csoonline.com, 03-Mar-2015. [Online]. Available: http://www.csoonline.com/article/2891834/mobile-security/geofencing-could-add-security-layer-for-mobile-devices.html?lf1=3501240615d618011202188d38053719. [Accessed: 22-Apr-2015].

[229] M. Williams, “Disruptive Technology on the Horizon”. [Online]. Available: http://www.govtech.com/products/Disruptive-Technology-on-the-Horizon.html. [Accessed: 06-May-2015].

[230] D. Storm, “Wearing GPS ankle monitor is a search, yet apps track your location every 3 minutes”, 01-Apr-2015. [Online]. Available: http://www.computerworld.com/article/2904833/wearing-gps-ankle-monitor-is-a-search-yet-apps-track-your-location-every-3-minutes.html. [Accessed: 20-May-2015].

[231] P. Aldhous and C. Seife, “See Maps Showing Where FBI Planes Are Watching From Above”. [Online]. Available: https://www.buzzfeed.com/peteraldhous/spies-in-the-skies?utm_term=.ntJykOvbG#.agqd2ylXZ. [Accessed: 13-Apr-2016].

[232] K. Hill, “You're being secretly tracked with facial recognition, even in church”. [Online]. Available: http://fusion.net/story/154199/facial-recognition-no-rules/. [Accessed: 06-Jan-2016].

[233] IEEE, “GM's Searching for Black Boxes, Next Time It Could Search the Cloud – IEEE Spectrum”, spectrum.ieee.org, 02-Jul-2014. [Online]. Available: http://spectrum.ieee.org/cars-that-think/transportation/safety/today-gm-seeks-legal-evidence-in-cars-black-boxestomorrow-lawyers-will-just-search-the-cloud/?utm_source=carsthatthink&utm_medium=email&u. [Accessed: 22-Apr-2015].

[234] K. Burnham, "Location Tracking: 6 Social App Settings To Check", http://www.informationweek.com/software/social/location-tracking-6-social-app-settings-to-check/d/d-id/1306643, 26-Apr-2014. [Online]. Available: http://www.informationweek.com/software/social/location-tracking-6-social-app-settings-to-check/d/d-id/1306643?page_number=1. [Accessed: 22-Apr-2015].

[235] M. Kumar, "Marauder's Map: Creepy Chrome extension that visualizes the location of your Facebook messages", 28-May-2015. [Online]. Available: http://geoawesomeness.com/marauders-map-creepy-chrome-extension-that-visualizes-the-location-of-your-facebook-messages/. [Accessed: 03-Jun-2015].

[236] H. Abelson, R. Anderson, S.M. Bellovin, J. Benaloh, M. Blaze, W. Diffie, J. Gilmore, M. Green, S. Landau, P.G. Neumann, R.L. Rivest, J.I. Schiller, B. Schneier, M. Specter and D. Weitzner, "Keys Under Doormats: Mandating insecurity by requiring government access to all data and communications", pp. 1–34, Jul. 2015.

[237] CISCO, "Industry Perspective: Understanding the Internet of Everything", 14-Feb-2015. [Online]. Available: https://afd34ee8b0806295b5a7-9fbee7de8d51db511b5de86d75069107.ssl.cf1.rackcdn.com/1_-_Convergence-the_first_step_to_IoE.pdf. [Accessed: 06-May-2015].

[238] S. Curtis, "Hackers take control of Jeep while it was driving using just a laptop and mobile phone", 2015. [Online]. Available: http://www.canberratimes.com.au/digital-life/consumer-security/security-experts-hack-into-moving-jeep-and-seize-control-of-car-20150721-gihmn3.html. [Accessed: 22-Jul-2015].

[239] K. Zetter, "Feds Say That Banned Researcher Commandeered a Plane", 15-May-2015. [Online]. Available: https://www.wired.com/2015/05/feds-say-banned-researcher-commandeered-plane/. [Accessed: 22-Jul-2015].

[240] FBI, "Chris-Roberts-Application-for-Search-Warrant", 18-Apr-2015. [Online]. Available: http://www.wired.com/wp-content/uploads/2015/05/Chris-Roberts-Application-for-Search-Warrant.pdf. [Accessed: 02-Sep-2015].

[241] Ben Moir, Touching tech-wearable technology. 2015.

[242] W. Eggers and J. Bellman, "The journey to government's digital transformation", pp. 1–44, Oct. 2015.

[243] Wikipedia, "Digital 5 tech-wearable technology – Wikipedia, the free encyclopedia". [Online]. Available: https://en.wikipedia.org/wiki/Digital_5. [Accessed: 03-Aug-2016].

APPENDICES

APPENDIX A: LIST OF EARTH OBSERVATION SATELLITES AS OF 30 JUNE 2016

(Source: Union of Concerned Scientists)

COSPAR Number	NORAD Number	Name of Satellite\Alternate Names	Country	Operator/Owner	Date of Launch
2016-025E	41460	AAUSat-4	Denmark	University of Aalborg	25/04/2016
1998-029A	25336	Advanced Orion 2 (NROL 6, USA 139)	USA	National Reconnaissance Office (NRO)	9/05/1998
2003-041A	27937	Advanced Orion 3 (NROL 19, USA 171)	USA	National Reconnaissance Office (NRO)	9/09/2003
2009-001A	33490	Advanced Orion 4 (NRO L-26, USA 202)	USA	National Reconnaissance Office (NRO)	18/01/2009
2010-063A	37232	Advanced Orion 5 (NRO L-32, USA 223)	USA	National Reconnaissance Office (NRO)	21/11/2010
2012-034A	38528	Advanced Orion 6 (NRO L-15, USA 237)	USA	National Reconnaissance Office (NRO)	29/06/2012
2016-036A	41584	Advanced Orion 7 (NRO L-37, USA 268)	USA	National Reconnaissance Office (NRO)	11/06/2016
2007-015A	31304	AIM (Aeronomy of Ice in Mesosphere)	USA	Center for Atmospheric Sciences, Hampton University/NASA	25/04/2007
2010-035D	36798	Alsat-2A (Algeria Satellite 2A)	Algeria	Centre National des Techniques Spatiales (CNTS)	12/07/2010
2014-070A	40298	ASNARO 1 (Advanced Satellite with New system Architecture for Observation)	Japan	Ministry of Economy, Trade and Industry	6/11/2014
2015-009A	40420	BARS-M (Cosmos 2503)	Russia	Ministry of Defense	27/02/2015
2016-020A	41394	BARS-M (Cosmos 2515)	Russia	Ministry of Defense	24/03/2016
2005-043A	28890	BeijinGalaxy-1 (Beijing 1 [Tsinghua], Tsinghau-2, China DMC+4)	China	Beijing Landview Mapping Information Technology Co. Ltd (BLMIT)	27/10/2005
2016-040F	41604	BIROS (Bispectral Infrared Optical System)	Germany	German Aerospace Center (DLR)	22/06/2016
2012-039B	38708	BKA 2 (BelKA 2)	Belarus	National Academy of Sciences	22/07/2012
2012-039A	38707	Canopus-B (Kanopus Vulcan 1)	Russia	Scientific Production Corporation (joint stock creation of Russian Space Agency)	22/07/2012
2005-017A	28649	CartoSat 1 (IRS P5)	India	Indian Space Research Organization (ISRO)	5/05/2005
2007-001B	29710	CartoSat 2 (IRS P7, CartoSat 2AT)	India	Indian Space Research Organization (ISRO)	10/01/2007
2008-021A	32783	CartoSat 2A	India	Indian Space Research Organization (ISRO)	28/04/2008
2010-035A	36795	CartoSat 2B	India	Indian Space Research Organization (ISRO)	12/07/2010
2016-040A	41599	CartoSat 2C	India	Indian Space Research Organization (ISRO)	22/06/2016
2014-079A	40336	CBERS 4 (China-Brazil Earth Resources Satellite 4)	China/Brazil	China National Space Administration (China)/National Institute for Space Research (Brazil)	7/12/2014
2008-056B	33434	Chuangxin 1-2 (Innovation 1-2)	China	Chinese Academy of Sciences	5/11/2008
2011-068A	37930	Chuangxin 1-3 (Innovation 1-3)	China	Chinese Academy of Sciences	20/11/2011
2014-051B	40137	Chuangxin 1-4 (Innovation 1-4)	China	Chinese Academy of Sciences	4/09/2014

2006-016A	29107	Cloudsat	USA	NASA/Colorado State University	28/04/2006
2014-084A	40353	Condor E2	South Africa	Armed Forces	19/12/2014
2006-011A	29047	COSMIC-A (Formosat-3A, Constellation Observing System for Meteorology, Ionosphere and Climate)	Taiwan/USA	National Space Program Office (NSPO)/University Corporation for Atmospheric Research (UCAR) Boulder, CO	15/04/2006
2006-011B	29048	COSMIC-B (Formosat-3B, Constellation Observing System for Meteorology, Ionosphere and Climate)	Taiwan/USA	National Space Program Office (NSPO)/University Corporation for Atmospheric Research (UCAR) Boulder, CO	15/04/2006
2006-011D	29050	COSMIC-D (Formosat-3D, Constellation Observing System for Meteorology, Ionosphere and Climate)	Taiwan/USA	National Space Program Office (NSPO)/University Corporation for Atmospheric Research (UCAR) Boulder, CO	15/04/2006
2006-011E	29051	COSMIC-E (Formosat-3E, Constellation Observing System for Meteorology, Ionosphere and Climate)	Taiwan/USA	National Space Program Office (NSPO)/University Corporation for Atmospheric Research (UCAR) Boulder, CO	15/04/2006
2006-011F	29052	COSMIC-F (Formosat-3F, Constellation Observing System for Meteorology, Ionosphere and Climate)	Taiwan/USA	National Space Program Office (NSPO)/University Corporation for Atmospheric Research (UCAR) Boulder, CO	15/04/2006
2007-023A	31598	COSMO-Skymed 1 (Constellation of small Satellites for Mediterranean basin Observation)	Italy	Italian Space Agency/Ministry of Defense	8/06/2007
2007-059A	32376	COSMO-Skymed 2 (Constellation of small Satellites for Mediterranean basin Observation)	Italy	Italian Space Agency/Ministry of Defense	9/12/2007
2008-054A	33412	COSMO-Skymed 3 (Constellation of small Satellites for Mediterranean basin Observation)	Italy	Italian Space Agency/Ministry of Defense	25/10/2008
2010-060A	37216	COSMO-Skymed 4 (Constellation of small Satellites for Mediterranean basin Observation)	Italy	Italian Space Agency/Ministry of Defense	6/11/2010
2015-003E	40380	CP-10 (Exocube)	USA	California Polytechnic State University/NASA JPL	31/01/2015
2010-013A	36508	Cryosat-2	ESA	European Space Agency (ESA)	8/04/2010
2014-029A	39766	Daichi-2 (Advanced Land Observing Satellite-2, ALOS 2 2)	Japan	Japan Aerospace Exploration Agency (JAXA)	24/05/2014
2009-041A	35681	Deimos 1	Spain	Deimos Imaging/DMC International Imaging (DMCII)	29/07/2009
2014-033D	40013	Deimos 2	Spain	Deimos Imaging/DMC International Imaging (DMCII)	19/06/2014
2015-032A	40715	DMC 3-1 (Disaster Monitoring Constellation)	United Kingdom	Surrey Satellite Technology Ltd.	10/07/2015
2015-032B	40716	DMC 3-2 (Disaster Monitoring Constellation)	United Kingdom	Surrey Satellite Technology Ltd.	10/07/2015
2015-032C	40717	DMC 3-3 (Disaster Monitoring Constellation)	United Kingdom	Surrey Satellite Technology Ltd.	10/07/2015
1997-012A	24753	DMSP 5D-2 F14 (Defense Meteorological Satellites Program, USA 131)	USA	DoD/NOAA	4/04/1997
1999-067A	25991	DMSP 5D-3 F15 (Defense Meteorological Satellites Program, USA 147)	USA	DoD/NOAA	12/12/1999

2003-048A	28054	DMSP 5D-3 F16 (Defense Meteorological Satellites Program, USA 172)	USA	DoD/NOAA	18/10/2003
2006-050A	29522	DMSP 5D-3 F17 (Defense Meteorological Satellites Program, USA 173)	USA	DoD/NOAA	4/11/2006
2009-057A	35951	DMSP 5D-3 F18 (Defense Meteorological Satellites Program, USA 210)	USA	DoD/NOAA	18/10/2009
2014-033T	40027	Dove 1c-1 (0 Flock 1C-1 0903)	USA	Planet Labs, Inc.	19/06/2014
2014-033P	40023	Dove 1c-10 (0 Flock 1C-10 090C)	USA	Planet Labs, Inc.	19/06/2014
2014-033Z	40033	Dove 1c-11 (0 Flock 1C-11 090D)	USA	Planet Labs, Inc.	19/06/2014
2014-033V	40029	Dove 1c-2 (0 Flock 1C-2 0904)	USA	Planet Labs, Inc.	19/06/2014
2014-033AH	40041	Dove 1c-3 (0 Flock 1C-3 0905)	USA	Planet Labs, Inc.	19/06/2014
2014-033X	40031	Dove 1c-4 (0 Flock 1C-4 0906)	USA	Planet Labs, Inc.	19/06/2014
2014-033AE	40038	Dove 1c-5 (0 Flock 1C-5 0907)	USA	Planet Labs, Inc.	19/06/2014
2014-033AC	40036	Dove 1c-6 (0 Flock 1C-6 0908)	USA	Planet Labs, Inc.	19/06/2014
2014-033S	40026	Dove 1c-7 90 Flock 1C-7 0909)	USA	Planet Labs, Inc.	19/06/2014
2014-033AG	40040	Dove 1c-8 (0 Flock 1C-8 090A)	USA	Planet Labs, Inc.	19/06/2014
2014-033AB	40035	Dove 1c-9 (0 Flock 1C-9 090B)	USA	Planet Labs, Inc.	19/06/2014
1998-067GP	40737	Dove 1e-10 (0 Flock 1E-10 0B09)	USA	Planet Labs, Inc.	15/07/2015
1998-067GQ	40738	Dove 1e-11 (0 Flock 1E-11 0B1E)	USA	Planet Labs, Inc.	15/07/2015
1998-067GR	40739	Dove 1e-12 (0 Flock 1E-12 0B0A)	USA	Planet Labs, Inc.	15/07/2015
1998-067GS	40740	Dove 1e-13 (0 Flock 1E-13 0C07)	USA	Planet Labs, Inc.	16/07/2015
1998-067GT	40741	Dove 1e-14 (0 Flock 1E-14 0B0F)	USA	Planet Labs, Inc.	16/07/2015
1998-067GM	40729	Dove 1e-6 (0 Flock 1E-6 0B0B)	USA	Planet Labs, Inc.	15/07/2015
1998-067GJ	40726	Dove 1e-7 (0 Flock 1E-7 0B0E)	USA	Planet Labs, Inc.	14/07/2015
1998-067HB	40950	Dove 2b-1 (0 Flock 2B-1 0C70)	USA	Planet Labs, Inc.	6/10/2015
1998-067HL	40963	Dove 2b-10 (0 Flock 2B-10 0C64)	USA	Planet Labs, Inc.	7/10/2015
1998-067HM	40979	Dove 2b-13 (0 Flock 2B-13 0C74)	USA	Planet Labs, Inc.	7/10/2015
1998-067HN	40980	Dove 2b-14 (0 Flock 2B-14 0C47)	USA	Planet Labs, Inc.	7/10/2015
1998-067HC	40951	Dove 2b-2 (0 Flock 2B-2 0C18)	USA	Planet Labs, Inc.	6/10/2015
1998-067HD	40952	Dove 2b-3 (0 Flock 2B-3 0C53)	USA	Planet Labs, Inc.	6/10/2015
1998-067HE	40953	Dove 2b-4 (0 Flock 2B-4 0C72)	USA	Planet Labs, Inc.	6/10/2015
1998-067HF	40954	Dove 2b-5 (0 Flock 2B-5 0C73)	USA	Planet Labs, Inc.	6/10/2015

1998-067HG	40955	Dove 2b-6 (0 Flock 2B-6 0C69)	USA	Planet Labs, Inc.	6/10/2015
1998-067HH	40956	Dove 2b-7 (0 Flock 2B-7 0C66)	USA	Planet Labs, Inc.	6/10/2015
1998-067HK	40962	Dove 2b-9 (0 Flock 2B-9 0C68)	USA	Planet Labs, Inc.	7/10/2015
1998-067JD	41483	Dove 2e-1 (0 Flock 2E-1 0C37)	USA	Planet Labs, Inc.	17/05/2016
1998-067JE	41484	Dove 2e-2 (0 Flock 2E-2 0C78)	USA	Planet Labs, Inc.	17/05/2016
1998-067JG	41486	Dove 2e-3 (0 Flock 2E-3 0C60)	USA	Planet Labs, Inc.	17/05/2016
1998-067JH	41487	Dove 2e-4 (0 Flock 2E-4 0C41)	USA	Planet Labs, Inc.	17/05/2016
1998-067HZ	41479	Dove 2ep-1 (0 Flock 2EP-1 0D05)	USA	Planet Labs, Inc.	17/05/2016
1998-067JB	41481	Dove 2ep-2 (0 Flock 2EP-2 0C1B)	USA	Planet Labs, Inc.	17/05/2016
1998-067JA	41480	Dove 2ep-3 (0 Flock 2EP-3 0D06)	USA	Planet Labs, Inc.	17/05/2016
1988-067JC	41482	Dove 2ep-4 (0 Flock 2EP-4 0C22)	USA	Planet Labs, Inc.	17/05/2016
2016-040U	41617	Dove 2p-1 ((0 Flock 2P-1 0E0D)	USA	Planet Labs, Inc.	22/06/2016
2016-040L	41609	Dove 2p-2 (0 Flock 2P-2 (0e0E)	USA	Planet Labs, Inc.	22/06/2016
2016-040V	41618	Dove 2p-3 (0 Flock 2P-3 (0E0F)	USA	Planet Labs, Inc.	22/06/2016
2016-040N	41611	Dove 2p-4 (0 Flock 2P-4 0E14)	USA	Planet Labs, Inc.	22/06/2016
2016-040T	41616	Dove 2p-5 (0 Flock 2P-5 0E16)	USA	Planet Labs, Inc.	22/06/2016
2016-040H	41606	Dove 2p-6 (0 Flock 2P-6 0E19)	USA	Planet Labs, Inc.	22/06/2016
2016-040S	41615	Dove 2p-7 (0 Flock 2P-7 0E1F)	USA	Planet Labs, Inc.	22/06/2016
2016-040Q	41613	Dove 2p-8 (0 Flock 2P-8 0E20)	USA	Planet Labs, Inc.	22/06/2016
2016-040M	41610	Dove 2p-9 (0 Flock 2P-9 0E26)	USA	Planet Labs, Inc.	22/06/2016
2016-040P	41612	Dove 2p-10 (0 Flock 2P-10 0E2F)	USA	Planet Labs, Inc.	22/06/2016
2016-040K	41608	Dove 2p-11 (0 Flock 2P-11 0E30)	USA	Planet Labs, Inc.	22/06/2016
2016-040R	41614	Dove 2p-12 (0 Flock 2P-12 0E3A)	USA	Planet Labs, Inc.	22/06/2016
1997-008A	24737	DSP 18 (USA 130) (Defense Support Program)	USA	Air Force	23/02/1997
2000-024A	26356	DSP 20 (USA 149) (Defense Support Program)	USA	Air Force	18/05/2000
2001-033A	26880	DSP 21 (USA 159) (Defense Support Program)	USA	Air Force	6/08/2001
2004-004A	28158	DSP 22 (USA 176) (Defense Support Program)	USA	Air Force	14/02/2004
2009-041B	35682	DubaiSat-1	United Arab Emirates	Emirates Institution for Advanced Science & Technology (EIAST)	29/07/2009
2013-066D	39419	DubaiSat-2	United Arab Emirates	Emirates Institution for Advanced Science & Technology (EIAST)	21/11/2013
2015-066A	41032	EKS-1 (Integrated Space System,	Russia	Ministry of Defense	17/11/2015

		Cosmos 2510, Tundra 11L)			
2011-001A	37344	Electro-L1 (GOMS 2 [Geostationary Operational Meteorological Satellite 2]	Russia	Roshydromet - Planeta	20/01/2011
2015-074A	41105	Electro-L2	Russia	Roshydromet - Planeta	11/12/2015
2011-076D	38010	ELISA-E12 (ELectronic Intelligence by SAtellite)	France	DGA (Arms Procurement Agency)/Centre National d'Etudes Spatiales (CNES)	17/12/2011
2011-076B	38008	ELISA-E24 (ELectronic Intelligence by SAtellite)	France	DGA (Arms Procurement Agency)/Centre National d'Etudes Spatiales (CNES)	17/12/2011
2011-076A	38007	ELISA-W11 (ELectronic Intelligence by SAtellite)	France	DGA (Arms Procurement Agency)/Centre National d'Etudes Spatiales (CNES)	17/12/2011
2011-076C	38009	ELISA-W23 (ELectronic Intelligence by SAtellite)	France	DGA (Arms Procurement Agency)/Centre National d'Etudes Spatiales (CNES)	17/12/2011
1999-068A	25994	EOS-AM Terra	USA/Canada/Japa n	Earth Sciences Enterprise (NASA)	18/12/1999
2004-026A	28376	EOS-CHEM Aura	USA	Goddard Space Flight Center/EOS Data and Operations System	15/07/2004
2002-022A	27424	EOS-PM Aqua (Advanced Microwave Scanning Radiometer for EOS, EOS PM-1)	USA/Japan/Brazil	NASA - Earth Science Enterprise/Japan Meteorological Agency/Brazilian Space Agency	4/05/2002
2000-079A	26631	EROS A1 (Earth Resources Observation Satellite)	Israel	ImageSat International, NV/Ministry of Defense	5/12/2000
2006-014A	29079	EROS B1 (Earth Resources Observation Satellite)	Israel	ImageSat International, NV/Ministry of Defense	25/04/2006
2006-053A	29640	Fengyun 2D (FY-2D)	China	China Meteorological Administration	8/12/2006
2008-066A	33463	Fengyun 2E (FY-2E)	China	China Meteorological Administration	23/12/2008
2012-002A	38049	Fengyun 2F (FY-2F)	China	China Meteorological Administration	12/01/2012
2014-090A	40367	Fengyun 2G (FY 2G)	China	China Meteorological Administration	31/12/2014
2008-026A	32958	Fengyun 3A (FY-3A)	China	China Meteorological Administration	27/05/2008
2010-059A	37214	Fengyun 3B (FY-3B)	China	China Meteorological Administration	4/11/2010
2013-052A	39260	Fengyun 3C (FY-3C)	China	China Meteorological Administration	23/09/2013
2010-046A	37162	FIA Radar 1 (Future Imagery Architecture (FIA) Radar 1, NROL-41, USA 215, Topaz)	USA	National Reconnaissance Office (NRO)	21/09/2010
2012-014A	38109	FIA Radar 2 (Future Imagery Architecture (FIA) Radar 2, NROL-25, USA 234, Topaz)	USA	National Reconnaissance Office (NRO)	3/04/2012
2013-072A	39462	FIA Radar 3 (Future Imagery Architecture (FIA) Radar 3, NROL-39 , USA 247, Topaz)	USA	National Reconnaissance Office (NRO)	6/12/2013
2016-010A	41334	FIA Radar 4 (Future Imagery Architecture (FIA) Radar 4, NROL-45, USA 267, Topax)	USA	National Reconnaissance Office (NRO)	10/02/2016

2004-018A	28254	Formosat-2 (ROCSAT-2, Republic of China Satellite 2)	Taiwan	National Space Program Office (NSPO)	21/05/2004
1997-047A	24920	FORTÉ (Fast On-orbit Recording of Transient Events)	USA	Los Alamos National Labs/DOE	29/08/1997
2013-018A	39150	Gaofen 1	China	Shanghai Academy of Spaceflight Technology (SAST)	26/04/2013
2014-049A	40118	Gaofen 2	China	Shanghai Academy of Spaceflight Technology (SAST)	19/08/2014
2015-083A	41194	Gaofen 4	China	China Aerospace Science and Technology Corporation (CASTC)	28/12/2015
2015-030A	40701	Gaofen 8	China	Shanghai Academy of Spaceflight Technology (SAST)	26/06/2015
2015-047A	40894	Gaofen 9	China	China Aerospace Science and Technology Corporation (CASTC)	14/09/2015
2008-042A	33331	GeoEye-1 (Orbview 5)	USA	DigitalGlobal Corporation	6/09/2008
2012-025A	38337	Global Change Observation Mission - 1 Water (GCOM-1, Shikuzu)	USA/Japan	Japan Aerospace Exploration Agency (JAXA)	17/05/2012
2006-018A	29155	GOES 13 (Geostationary Operational Environmental Satellite, GOES-N)	USA	NOAA (National Oceanographic and Atmospheric Administration)	24/05/2006
2009-033A	35491	GOES 14 (Geostationary Operational Environmental Satellite, GOES-O)	USA	NOAA (National Oceanographic and Atmospheric Administration)	27/06/2009
2010-008A	36411	GOES 15 (Geostationary Operational Ennvironmental Satellite, GOES-P)	USA	NOAA (National Oceanographic and Atmospheric Administration)	4/03/2010
2012-073A	39030	Göktürk 2	Turkey	Turkish Ministry of National Defense	18/12/2012
2014-009C	39574	GPM Core Observatory (Global Precipitation Measurement)	USA/Japan	NASA/JAXA	27/02/2014
2002-012A	27391	Grace 1 (Gravity Recovery and Climate Experiment, "Tom and Jerry")	Germany/USA	GeoForschungsZentrum (GFZ)/Center for Space Research, University of Texas	17/03/2002
2002-012B	27392	Grace 2 (Gravity Recovery and Climate Experiment, "Tom and Jerry")	Germany/USA	GeoForschungsZentrum (GFZ)/Center for Space Research, University of Texas	17/03/2002
2009-002A	33492	Greenhouse Gases Observing Satellite (Ibuki, GoSAT)	Japan	Japan Aerospace Exploration Agency (JAXA)	23/01/2009
2011-043A	37781	Haiyang 2A (HY 2A)	China	State Oceanic Administration (SOA)	15/08/2011
2004-049A	28492	Helios 2A	France/Italy/ Belgium/Spain/ Greece	Centre National d'Etudes Spatiales (CNES)/Délégation Générale de l'Armement (DGA)	18/12/2004
2009-073A	36124	Helios 2B	France/Italy/ Belgium/Spain/ Greece	Centre National d'Etudes Spatiales (CNES)/Délégation Générale de l'Armement (DGA)	18/12/2009
2014-060A	40267	Himawari 8	Japan	Japan Meteorological Agency/Meteorological Satellite Center (MSC)	7/10/2014
2008-041A	33320	HJ-1A (Huan Jing 1A)	China	National Remote Sensing Center (NRSCC)	5/09/2008
2008-041B	33321	HJ-1B (Huan Jing 1B)	China	National Remote Sensing Center (NRSCC)	5/09/2008
2012-064A	38997	HJ-1C (Huan Jing 1C)	China	National Committee for Disaster Reduction and State Environmental Protection	18/11/2012

2014-070B	40299	Hodoyoshi-1	Japan	University of Tokyo and NESTRA	6/11/2014
2006-037A	29393	IGS-3A (Information Gathering Satellite 3A, IGS Optical 2)	Japan	Cabinet Satellite Intelligence Center (CSIC)	11/09/2006
2009-066A	36104	IGS-5A (Information Gathering Satellite 5A, IGS Optical 3)	Japan	Cabinet Satellite Intelligence Center (CSIC)	28/11/2009
2011-050A	37813	IGS-6A (Information Gathering Satellite 6A, IGS Optical 4)	Japan	Cabinet Satellite Intelligence Center (CSIC)	23/09/2011
2011-075A	37954	IGS-7A (Information Gathering Satellite 7A, IGS Radar 3)	Japan	Cabinet Satellite Intelligence Center (CSIC)	12/12/2011
2013-002A	39061	IGS-8A (Information Gathering Satellite 8A, IGS Radar 4)	Japan	Cabinet Satellite Intelligence Center (CSIC)	27/01/2013
2013-002B	39062	IGS-8B (Information Gathering Satellite 8B, IGS Optical Demonstrator)	Japan	Cabinet Satellite Intelligence Center (CSIC)	27/01/2013
2015-004A	40381	IGS-9A (Information Gathering Satellite 9A, IGS Radar 5)	Japan	Cabinet Satellite Intelligence Center (CSIC)	31/01/2015
2015-015A	40538	IGS-9B (Information Gathering Satellite 9B, IGS Optical 5)	Japan	Cabinet Satellite Intelligence Center (CSIC)	26/03/2015
2006-027A	29249	Improved Trumpet 4 (NROL-22, National Reconnaissance Office Launch-22, SBIRS HEO-1, Twins 1, USA 184)	USA	National Reconnaissance Office (NRO)	28/06/2006
2008-010A	32706	Improved Trumpet 5 (NROL-28, National Reconnaissance Office Launch-28, SBIRS HEO-2, Twins 2, USA 200)	USA	National Reconnaissance Office (NRO)	13/03/2008
2014-081A	40344	Improved Trumpet 6 (NROL-35, National Reconnaissance Office Launch-35, SBIRS HEO-3, USA 259)	USA	National Reconnaissance Office (NRO)	13/12/2014
2003-013A	27714	INSAT 3A (Indian National Satellite)	India	Indian Space Research Organization (ISRO)	9/04/2003
2013-038B	39216	INSAT 3D (Indian National Satellite)	India	Indian Space Research Organization (ISRO)	25/07/2013
2003-046A	28051	IRS-P6 (Resourcesat-1)	India	Indian Space Research Organization (ISRO)	17/10/2003
2008-032A	33105	Jason 2	USA/France	NASA/Centre National d'Etudes Spatiales (CNES)/NOAA/EUMETSAT	20/06/2008
2016-002A	41240	Jason 3	USA/France	NASA/Centre National d'Etudes Spatiales (CNES)/NOAA/EUMETSAT	17/01/2016
2015-057D	40961	Jilin-1	China	Changchun Institute of Optics	7/10/2015
2011-058B	37839	Jugnu	India	Indian Institute of Technology Kanpur	12/10/2011
2002-043A	27525	Kalpana-1 (Metsat-1)	India	Indian Space Research Organization (ISRO)	12/09/2002
2014-024A	39731	KazEOSat-1 (Kazcosmos Earth Observation Satellite)	Kazakhstan	Kazcosmos	30/04/2014
2014-033A	40010	KazEOSat-2 (kazcosmos Earth Observation Satellite)	Kazakhstan	Kazcosmos	19/06/2014
2015-077B	41167	Kent Ridge 1	Singapore	National University of Singapore	16/12/2015
2005-042A	28888	Keyhole 5 (Advanced KH-11, KH-12-5, Improved Crystal, EIS-3, USA 186)	USA	National Reconnaissance Office (NRO)	19/10/2005
2011-002A	37348	Keyhole 6 (NRO L49, Advanced KH-11, KH-12-6, Improved Crystal, USA 224)	USA	National Reconnaissance Office (NRO)	20/01/2011

2013-043A	39232	Keyhole 7 (NRO L65, Advanced KH-11, Improved Crystal, USA 245)	USA	National Reconnaissance Office (NRO)	28/08/2013
2012-025B	38338	Kompsat-3 (Arirang 3, Korean Multipurpose Satellite-3)	South Korea	Korea Aerospace Research Institute (KARI)	17/05/2012
2015-014A	40536	Kompsat-3A (Arirang 3A, Korean Multipurpose Satellite-3A)	South Korea	Korea Aerospace Research Institute (KARI)	25/03/2015
2013-042A	39227	Kompsat-5 (Arirang 5, Korean Multipurpose Satellite-4)	South Korea	Korea Aerospace Research Institute (KARI)	22/08/2013
2013-032A	39194	Kondor (Cosmos 2487)	Russia	Ministry of Defense	27/06/2013
2014-073A	40311	Kuaizhou-2 (KZ-2)	China	National Academy of Sciences	21/11/2014
1997-064A	25017	Lacrosse/Onyx 3 (Lacrosse-3, USA 133)	USA	National Reconnaissance Office (NRO)	24/10/1997
2000-047A	26473	Lacrosse/Onyx 4 (Lacrosse-4, USA 152)	USA	National Reconnaissance Office (NRO)	17/08/2000
2005-016A	28646	Lacrosse/Onyx 5 (Lacrosse-5, NROL 16, USA 182)	USA	National Reconnaissance Office (NRO)	30/04/2005
1999-020A	25682	Landsat 7	USA	NASA/US Geological Survey	15/04/1999
2013-008A	39084	Landsat 8	USA	NASA/US Geological Survey	11/02/2013
2015-052B	40931	LAPAN A2	Indonesia	Indonesian National Aeronautics and Space Agency (Lembaga Penerbangan dan Antariksa Nasional - LAPAN)	24/09/2015
2016-040E	41603	LAPAN A3	Indonesia	Indonesian National Aeronautics and Space Agency (Lembaga Penerbangan dan Antariksa Nasional - LAPAN)	22/06/2016
2015-052G	40936	Lemur-2 Chris (Lemur-2 F4)	USA	Spire Global Inc.	28/09/2015
2015-052F	40935	Lemur-2 Jeroen (Lemur-2 F3)	USA	Spire Global Inc.	28/09/2015
2015-052D	40933	Lemur-2 Joel (Lemur-2 F1)	USA	Spire Global Inc.	28/09/2015
2015-052E	40934	Lemur-2 Peter (Lemur-2 F2)	USA	Spire Global Inc.	28/09/2015
2015-057B	40959	Lingqiao-A (Lingqiao Video A)	China	Changchun Institute of Optics	7/10/2015
2015-057C	40960	Lingqiao-B (Lingqiao Video B)	China	Changchun Institute of Optics	7/10/2015
2009-063A	36095	Lotos-S1 (Cosmos 2455)	Russia	Ministry of Defense	20/11/2009
2014-086A	40358	Lotos-S2 (Cosmos 2502)	Russia	Ministry of Defense	23/12/2014
2015-057A	40958	LQSat (Lingqiao Satellite, Jilian 1A)	China	Changchun Institute of Optics	7/10/2015
2016-040G	41605	M3MSat (Maritime Monitoring and Messaging Microsatellite)	Canada	Defence Research and Development Canada (DRDC)/Canadian Space Agency	22/06/2016
2011-058A	37838	Megha-Tropiques	India/France	Indian Space Research Organization (ISRO)/Centre National d'Etudes Spatiales (CNES)	12/10/2011
1994-054A	23223	Mercury 1 (Advanced Vortex 1, USA 105)	USA	National Reconnaissance Office (NRO)/USAF	27/08/1994
1996-026A	23855	Mercury 2 (Advanced Vortex 2, USA	USA	National Reconnaissance Office	24/04/1996

		118)		(NRO)/USAF	
2014-020A	39652	Mercury 3 (NROL 67, USA 250)	USA	National Reconnaissance Office (NRO)/USAF	10/04/2014
2014-037A	40069	Meteor-M N-2	Russia	Russian Federal Service For Hydrometeorology and Environmental Monitoring (ROSHYDROMET)	8/07/2014
2012-035B	38552	Meteosat 10 (MSGalaxy-3,MSG 3)	Multinational	EUMETSAT (European Organization for the Exploitation of Meteorological Satellites)	5/07/2012
2015-034A	40732	Meteosat 11 (MSG 4)	Multinational	EUMETSAT (European Organization for the Exploitation of Meteorological Satellites)	15/07/2015
1997-049B	24932	Meteosat 7 (MTP 1)	Multinational	EUMETSAT (European Organization for the Exploitation of Meteorological Satellites)	2/09/1997
2002-040B	27509	Meteosat 8 (MSGalaxy-1, MSG-1)	Multinational	EUMETSAT (European Organization for the Exploitation of Meteorological Satellites)	21/08/2002
2005-049B	28912	Meteosat 9 (MSGalaxy-2, MSG 2)	Multinational	EUMETSAT (European Organization for the Exploitation of Meteorological Satellites)	21/12/2005
2006-044A	29499	MetOp-A (Meteorological Operational satellite)	Multinational	ESA/EUMETSAT (European Organization for the Exploitation of Meterological Satellites)	19/10/2006
2012-049A	38771	MetOp-B (Meteorological Operational satellite)	Multinational	ESA/EUMETSAT (European Organization for the Exploitation of Meterological Satellites)	17/09/2012
2012-039E	38711	MKA-FKI-1 (Zond PP)	Russia	Lavochkin Research and Production Association (Lavochkin NPO)	22/07/2012
2006-004A	28937	MTSAT-2 (Multi-Functional Transport Satellite)	Japan	Japan Meteorological Agency/Meteorological Satellite Center (MSC)	18/02/2006
2011-044B	37789	NigeriaSat-2	Nigeria	National Space Research and Development Agency (NASRDA)	17/08/2011
1998-030A	25338	NOAA-15 (NOAA-K)	USA	National Oceanographic and Atmospheric Administration (NOAA) (part of international program)	13/05/1998
2005-018A	28654	NOAA-18 (NOAA-N, COSPAS-SARSAT)	USA	National Oceanographic and Atmospheric Administration (NOAA) (part of international program)	20/05/2005
2009-005A	33591	NOAA-19 (NOAA-N Prime, COSPAS-SARSAT)	USA	National Oceanographic and Atmospheric Administration (NOAA) (part of international program)	6/02/2009
2011-061A	37849	NPP (National Polar-orbiting Operational Environmental Satellite System [NPOESS])	USA	National Oceanographic and Atmospheric Administration (NOAA)/NASA	28/10/2011
2016-033B	41557	ÑuSat-1 (Fresco)	Argentina	Satellogic	29/05/2016
2016-033C	41558	ÑuSat-2 (Batata)	Argentina	Satellogic	29/05/2016
2009-051A	35931	Oceansat-2	India	Indian Space Research Organization (ISRO)	23/09/2009

2014-035A	40059	OCO 2 (Orbiting Carbon Observatory)	USA	National Aeronautics and Space Administration (NASA)	2/07/2014
2001-007A	26702	Odin	Sweden	Swedish National Space Board	20/02/2001
2014-019A	39650	Ofeq 10	Israel	Ministry of Defense	9/04/2014
2002-025A	27434	Ofeq 5	Israel	Ministry of Defense	28/05/2002
2007-025A	31601	Ofeq 7	Israel	Ministry of Defense	10/06/2007
2010-031A	36608	Ofeq 9	Israel	Ministry of Defense	22/06/2010
2013-028A	39177	Persona-2 (Cosmos 2486)	Russia	Ministry of Defense	7/06/2013
2015-029A	40699	Persona-3 (Cosmos 2506)	Russia	Ministry of Defense	23/06/2015
2011-076F	38012	Pléiades HR1A	France/Italy	Ministry of Defense/Centre National d'Etudes Spatiales (CNES) - cooperation with Austria, Belgium, Spain, Sweden	17/12/2011
2012-068A	39019	Pléiades HR1B	France	Ministry of Defense/Centre National d'Etudes Spatiales (CNES) - cooperation with Austria, Belgium, Spain, Sweden	2/12/2012
2009-002B	33493	Prism (Pico-satellite for Remote-sensing and Innovative Space Missions, Hitomi)	Japan	University of Tokyo	23/01/2009
2013-021A	39159	Proba V (Project for On-Board Autonomy)	ESA	European Space Agency (ESA)	7/05/2013
2014-070D	40301	Qsat-EOS (KYUshu SATellite - Earth Observation System)	Japan	Kyushu University	6/11/2014
2007-061A	32382	Radarsat-2	Canada	Radarsat International	14/12/2007
2008-040C	33314	RapidEye-1 (RapidEye-C)	Germany	RapidEye AG	29/08/2008
2008-040A	33312	RapidEye-2 (RapidEye A)	Germany	RapidEye AG	29/08/2008
2008-040D	33315	RapidEye-3 (RapidEye D)	Germany	RapidEye AG	29/08/2008
2008-040E	33316	RapidEye-4 (RapidEye E)	Germany	RapidEye AG	29/08/2008
2008-040B	33313	RapidEye-5 (RapidEye B)	Germany	RapidEye AG	29/08/2008
2011-044D	37791	RASAT	Turkey	Space Technologies Research Institute	17/08/2011
2014-037B	40070	Relek (ICA-FC1)	Russia	Skobeltsyn Institute of Nuclear Physics	8/07/2014
2011-015A	37387	Resourcesat 2 (exactView-2)	India/Canada	Indian Space Research Organization (ISRO)/exactEarth	20/04/2011
2006-021A	29228	Resurs-DK1 (Resurs - High Resolution 1)	Russia	TsSKB Progress (State Research & Production Space Rocket Center)	15/06/2006
2013-030A	39186	Resurs-P1	Russia	Russian Federal Space Agency (Roskosmos)	25/06/2013
2016-016A	41386	Resurs-P3	Russia	Russian Federal Space Agency (Roskosmos)	13/03/2016
2012-017A	38248	RISat-1 (Radar Imaging Satellite 1)	India	Ministry of Defense	25/04/2012
2009-019A	34807	RISat-2 (Radar Imaging Satellite 2)	India	Ministry of Defense	20/04/2009

2014-029D	39769	Rising-2	Japan	Tohoku University/Hokkaido University	24/05/2014
2013-009A	39086	SARAL (Satellite with ARGOS and ALTIKA)	India/France	Indian Space Research Organization (ISRO)/Centre National d'Etudes Spatiales (CNES)	25/02/2013
2006-060A	29658	SAR-Lupe 1	Germany	Armed Forces	19/12/2006
2007-030A	31797	SAR-Lupe 2	Germany	Armed Forces	2/07/2007
2007-053A	32283	SAR-Lupe 3	Germany	Armed Forces	1/11/2007
2008-014A	32750	SAR-Lupe 4	Germany	Armed Forces	27/03/2008
2008-036A	33244	SAR-Lupe 5	Germany	Armed Forces	22/07/2008
2004-025F	28371	Saudisat-2	Saudi Arabia	Riyadh Space Research Institute	29/06/2004
2007-012B	31118	Saudisat-3	Saudi Arabia	Riyadh Space Research Institute	17/04/2007
2011-019A	37481	SBIRS GEO 1 (Space Based Infrared System Geosynchronous 1, USA 230)	USA	US Air Force	7/05/2011
2013-011A	39120	SBIRS GEO 2 (Space Based Infrared System Geosynchronous 2, USA 241)	USA	US Air Force	19/03/2013
2005-004A	28537	SB-WASS 3-3 (Space Based Wide Area Surveillance System) (NOSS 3-3, USA 181, NRO L23)	USA	National Reconnaissance Office/US Navy	3/02/2005
2005-004B	28541	SB-WASS 3-3 (Space Based Wide Area Surveillance System) (NOSS 3-3, USA 181, NRO L23)	USA	National Reconnaissance Office/US Navy	3/02/2005
2007-027A	31701	SB-WASS 3-4 (Space Based Wide Area Surveillance System) NOSS 3-4, USA 194, NRO L30)	USA	National Reconnaissance Office/US Navy	15/06/2007
2007-027B	31708	SB-WASS 3-4 (Space Based Wide Area Surveillance System) NOSS 3-4, USA 194, NRO L30)	USA	National Reconnaissance Office/US Navy	15/06/2007
2011-014A	37386	SB-WASS 3-5 (Space Based Wide Area Surveillance System) NOSS 3-5, USA 229, NRO L34)	USA	National Reconnaissance Office/US Navy	15/04/2011
2011-014B	37391	SB-WASS 3-5 (Space Based Wide Area Surveillance System) NOSS 3-5, USA 229, NRO L34)	USA	National Reconnaissance Office/US Navy	15/04/2011
2012-048A	38758	SB-WASS 3-6 (Space Based Wide Area Surveillance System) NOSS 3-6, USA 238, NRO L36)	USA	National Reconnaissance Office/US Navy	13/09/2012
2012-048P	38773	SB-WASS 3-6 (Space Based Wide Area Surveillance System) NOSS 3-6, USA 238, NRO L36)	USA	National Reconnaissance Office/US Navy	13/09/2012
2015-058A	40964	SB-WASS 3-7 (Space Based Wide Area Surveillance System) NOSS 3-7, USA 264, NRO L55)	USA	National Reconnaissance Office/US Navy	8/10/2015
2015-058R	40981	SB-WASS 3-7 (Space Based Wide Area Surveillance System) NOSS 3-7, USA 264, NRO L55)	USA	National Reconnaissance Office/US Navy	8/10/2015
1993-009B	22491	SCD-1 (Satélite de Coleta de Dados)	Brazil	Instituto Nacional de Pesquisas Espaciais (INPE)	9/02/1993
1998-060A	25504	SCD-2 (Satélite de Coleta de Dados)	Brazil	Instituto Nacional de Pesquisas Espaciais (INPE)	23/10/1998

2014-016A	39634	Sentinel 1A	ESA	EUMETSAT	3/04/2014
2016-025A	41456	Sentinel 1B	ESA	EUMETSAT	25/04/2016
2015-028A	40697	Sentinel 2A	ESA	EUMETSAT	23/06/2015
2016-011A	41335	Sentinel 3A	ESA	EUMETSAT	16/02/2016
2010-051A	37179	Shijian 6G (SJ6-04A)	China	Chinese Academy of Space Technology (CAST)	6/10/2010
2010-051B	37180	Shijian 6H (SJ6_04B)	China	Chinese Academy of Space Technology (CAST)	6/10/2010
2004-012A	28220	Shiyan 1 (SY 1, Tansuo 1, Experimental Satellite 1)	China	Chinese Academy of Space Technology (CAST)	18/04/2004
2008-056A	33433	Shiyan 3 (SY3, Experimental Satellite 3)	China	Chinese Academy of Space Technology (CAST)	5/11/2008
2011-068B	37931	Shiyan 4 (SY4, Experimental Satellite 4)	China	Chinese Academy of Space Technology (CAST)	20/11/2011
2013-068A	39455	Shiyan 5 (SY5, Experimental Satellite 5)	China	Chinese Academy of Space Technology (CAST)	25/11/2013
2013-066C	39418	SkySat-1	USA	Terra Bella	21/11/2013
2014-037D	40072	SkySat-2	USA	Terra Bella	8/07/2014
2016-040C	41601	SkySat-3 (SkySat Gen 2-1)	USA	Terra Bella	22/06/2016
2015-003A	40376	SMAP (Soil Moisture Active Passive Satellite)	USA	National Aeronautics and Space Administration (NASA)	31/01/2015
2009-059A	36036	SMOS (Soil Moisture and Ocean Salinity satellite)	ESA	Centre National d'Etudes Spatiales (CNES)/European Space Agency	2/11/2009
2012-047A	38755	Spot 6 (Système Probatoire d'Observation de la Terre)	France/Belgium/Sweden	Spot Image	9/09/2012
2014-034A	40053	Spot 7 (Système Probatoire d'Observation de la Terre)	France/Belgium/Sweden	Spot Image	30/06/2014
2011-076E	38011	SSOT (Sistema Satelital para la Observación de la Tierra)	Chile	Chilean Air Force	17/12/2011
2013-067B	39452	SWARM-A	ESA	European Space Agency (ESA)	22/11/2013
2013-067A	39451	SWARM-B	ESA	European Space Agency (ESA)	22/11/2013
2013-067C	39453	SWARM-C	ESA	European Space Agency (ESA)	22/11/2013
2010-030A	36605	TanDEM-X (TerraSAR-X add-on for Digital Elevation Measurement)	Germany	German Aerospace Center (DLR)/Astrium	21/06/2010
2008-002A	32476	TecSAR (Ofeq 8, Polaris)	Israel	Defense Ministry	21/01/2008
2015-077D	41169	TeLEOS 1	Singapore	AgilSpace	16/12/2015
2007-026A	31698	TerraSAR-X 1 (Terra Synthetic Aperture Radar X-Band)	Germany	German Aerospace Center (DLR)/Infoterra	15/06/2007
2008-049A	33396	THEOS (Thailand Earth Observation System)	Thailand	Geo-Informatics and Space Technology Development Agency (GISTDA)	1/10/2008
2010-040A	36985	Tianhui 1-01	China	China Aerospace Science and Technology Corporation (CASTC)	24/08/2010

2012-020A	38256	Tianhui 1-02	China	China Aerospace Science and Technology Corporation (CASTC)	6/05/2012
2015-061A	40988	Tianhui 1-03	China	China Aerospace Science and Technology Corporation (CASTC)	26/10/2015
2014-033AK	40043	TIGRISat	Iraq	La Sapienza University of Rome	19/06/2014
1997-068A	25034	Trumpet 3 (NROL-4, National Reconnaissance Office Launch-4, USA 136)	USA	National Reconnaissance Office (NRO)/USAF	8/11/1997
2007-029A	31792	Tselina-2 (Cosmos 2428)	Russia	Ministry of Defense	29/06/2007
2014-070E	40302	TSUBAME	Japan	Tokyo Institute of Technology, Tokyo University of Science and JAXA	6/11/2014
2009-041C	35683	UK-DMC-2 (BNSCSat-2, British National Science Center Satellite 2)	United Kingdom	UK/DMC International Imaging (DMCII)	29/07/2009
2014-029B	39767	UNIFORM 1 (UNiversity International FORmation Mission 1)	Japan	Wakayama University (UNIFORM consortium)	24/05/2014
2012-046A	38752	Van Allen Probe A (RBSP-A, Radiation Belt Storm Probes)	USA	NASA/Johns Hopkins University Applied Physics Laboratory	30/08/2012
2012-046B	38753	Van Allen Probe B (RBSP-B, Radiation Belt Storm Probes)	USA	NASA/Johns Hopkins University Applied Physics Laboratory	30/08/2012
2013-021B	39160	VNREDSat 1A (Vietnam Natural Resources Environment and Disaster monitoring small Satellite)	Vietnam	Space Technology Institute-Vietnam Academy of Science and Technology (STI-VAST)	7/05/2013
2012-052A	38782	VRSS-1 (Venezuelan Remote Sensing Satellite, Francisco Miranda)	Venezuela	Bolivarian Agency for Space Activities	28/09/2012
2013-066H	39423	WNISat-1 Weather News Inc. Satellite 1)	Japan	Weathernews, Inc.	21/11/2013
2007-041A	32060	Worldview 1	USA	DigitalGlobe Corporation	18/09/2007
2009-055A	35946	Worldview 2	USA	DigitalGlobe Corporation	8/10/2009
2014-048A	40115	Worldview 3	USA	DigitalGlobe Corporation	13/08/2014
2011-015C	37389	X-Sat	Singapore	Centre for Research in Satellite Technology (CREST)	20/04/2011
2010-038A	36834	Yaogan 10 (Remote Sensing Satellite 10)	China	People's Liberation Army (C41)	9/08/2010
2010-047A	37165	Yaogan 11 (Remote Sensing Satellite 11)	China	People's Liberation Army (C41)	22/09/2010
2011-066B	37875	Yaogan 12 (Remote Sensing Satellite 12)	China	People's Liberation Army (C41)	9/11/2011
2011-072A	37941	Yaogan 13 (Remote Sensing Satellite 13)	China	People's Liberation Army (C41)	29/11/2011
2012-021A	38257	Yaogan 14 (Remote Sensing Satellite 14)	China	People's Liberation Army (C41)	10/05/2012
2012-029A	38354	Yaogan 15 (Remote Sensing Satellite 15)	China	People's Liberation Army (C41)	29/05/2012
2012-066A	39011	Yaogan 16A (Remote Sensing Satellite 16A, Yaogan Weixing 16)	China	People's Liberation Army (C41)	25/11/2012
2012-066B	39012	Yaogan 16B (Remote Sensing Satellite 16B)	China	People's Liberation Army (C41)	25/11/2012
2012-066C	39013	Yaogan 16C (Remote Sensing Satlelite 16C)	China	People's Liberation Army (C41)	25/11/2012

2013-046A	39239	Yaogan 17A (Remote Sensing Satellite 17A, Yaogan Weixing 17)	China	People's Liberation Army (C41)	1/09/2013
2013-046B	39240	Yaogan 17B (Remote Sensing Satellite 17B)	China	People's Liberation Army (C41)	1/09/2013
2013-046C	39241	Yaogan 17C (Remote Sensing Satellite 17C)	China	People's Liberation Army (C41)	1/09/2013
2013-059A	39363	Yaogan 18 (Remote Sensing Satellite 18)	China	People's Liberation Army (C41)	29/10/2013
2013-065A	39410	Yaogan 19 (Remote Sensing Satellite 19)	China	People's Liberation Army (C41)	20/11/2013
2014-047A	40109	Yaogan 20A (Remote Sensing Satellite 20A)	China	People's Liberation Army (C41)	9/08/2014
2014-047B	40110	Yaogan 20B (Remote Sensing Satellite 20B)	China	People's Liberation Army (C41)	9/08/2014
2014-047C	40111	Yaogan 20C (Remote Sensing Satellite 20C)	China	People's Liberation Army (C41)	9/08/2014
2014-053A	40143	Yaogan 21 (Remote Sensing Satellite 21)	China	People's Liberation Army (C41)	8/09/2014
2014-063A	40275	Yaogan 22 (Remote Sensing Satellite 22)	China	People's Liberation Army (C41)	20/10/2014
2014-071A	40305	Yaogan 23 (Remote Sensing Satellite 23)	China	People's Liberation Army (C41)	14/11/2014
2014-072A	40310	Yaogan 24 (Remote Sensing Satellite 24)	China	People's Liberation Army (C41)	20/11/2014
2014-080A	40338	Yaogan 25A (Remote Sensing Satellite 25A)	China	People's Liberation Army (C41)	10/12/2014
2014-080B	40339	Yaogan 25B (Remote Sensing Satellite 25B)	China	People's Liberation Army (C41)	10/12/2014
2014-0808C	40340	Yaogan 25C (Remote Sensing Satellite 25C)	China	People's Liberation Army (C41)	10/12/2014
2014-088A	40362	Yaogan 26 (Remote Sensing Satellite 26)	China	People's Liberation Army (C41)	27/12/2014
2015-040A	40878	Yaogan 27 (Remote Sensing Satellite 27)	China	People's Liberation Army (C41)	27/08/2015
2015-064A	41026	Yaogan 28 (Remote Sensing Satellite 28)	China	People's Liberation Army (C41)	8/11/2015
2015-069A	41038	Yaogan 29 (Remote Sensing Satellite 29)	China	People's Liberation Army (C41)	26/11/2015
2016-029A	41473	Yaogan 30 (Remote Sensing Satellite 30)	China	People's Liberation Army (C41)	15/05/2016
2008-061A	33446	Yaogan 4 (Remote Sensing Satellite 4)	China	People's Liberation Army (C41)	1/12/2008
2009-021A	34839	Yaogan 6 (Remote Sensing Satellite 6, Jian Bing 7-A)	China	People's Liberation Army (C41)	22/04/2009
2009-069A	36110	Yaogan 7 (Remote Sensing Satellite 7)	China	People's Liberation Army (C41)	9/12/2009
2009-072A	36121	Yaogan 8 (Remote Sensing Satellite 8)	China	People's Liberation Army (C41)	15/12/2009
2010-009A	36413	Yaogan 9A (Remote Sensing Satellite 9A)	China	People's Liberation Army (C41)	5/03/2010
2010-009B	36414	Yaogan 9B (Remote Sensing Satellite 9B)	China	People's Liberation Army (C41)	5/03/2010
2010-009C	36415	Yaogan 9C (Remote Sensing Satellite 9C)	China	People's Liberation Army (C41)	5/03/2010

2004-044A	28470	Zhangguo Ziyuan 2C (ZY-2C, JB-3C)	China	Chinese Academy of Space Technology	4/11/2004
2011-079A	38038	Ziyuan 1-02C	China	China Centre for Resources Satellite Data and Application (CRESDA)	22/12/2011
2012-001A	38046	Ziyuan 3 (ZY-3)	China	China Centre for Resources Satellite Data and Application (CRESDA)	9/01/2012
2016-033A	41556	Ziyuan 3-2	China	China Centre for Resources Satellite Data and Application (CRESDA)	29/05/2016
2010-032A	36744	COMS-1 (Communication, Ocean and Meteorological Satellite; Cheollian)	South Korea	Korea Aerospace Research Institute (KARI)	26/06/2010
2014-053B	40144	Tiantuo-2	China	National University of Defense Technology	8/09/2014
2014-087A	40360	Resurs-P2	Russia	Russian Federal Space Agency (Roskosmos)	26/12/2014
2000-075A	26619	EO-1 (Earth Observing 1)	USA	NASA Earth Science Office	21/11/2000
2001-049C	26959	Bird 2 (Bispectral InfraRed Detector 2)	Germany	Institute of Space Sensor Technology and Planetary Exploration	22/10/2001
2015-058E	40968	BisonSat (Nwist Qwiqway)	USA	Salish Kootenai College	8/10/2015
2006-031A	29268	Kompsat-2 (Arirang 2, Korean planned Multipurpose Satellite-2)	South Korea	Korea Aerospace Research Institute (KARI)	28/07/2006
2007-001A	29709	LAPAN-Tubsat (LAPAN A1)	Indonesia	Indonesian National Aeronautics and Space Agency (Lembaga Penerbangan dan Antariksa Nasional - LAPAN)	10/01/2007
2011-044C	37790	Nigeriasat-X	Nigeria	National Space Research and Development Agency (NASRDA)	17/08/2011

APPENDIX B: LIST OF NAVIGATION AND GLOBAL POSITIONING SATELLITES AS OF 30 JUNE 2016

(Source: Union of Concerned Scientists)

COSPAR	NORAD	Name of Satellite, Alternate Names	Country	Operator/Owner	Date of Launch
2012-018A	38250	Beidou 2-12 (Compass M3)	China	Chinese Defense Ministry	28/04/2012
2012-018B	38251	Beidou 2-13 (Compass M4)	China	Chinese Defense Ministry	28/04/2012
2012-050A	38774	Beidou 2-14 (Compass M5)	China	Chinese Defense Ministry	18/09/2012
2012-050B	38775	Beidou 2-15 (Compass M6)	China	Chinese Defense Ministry	18/09/2012
2012-059A	38953	Beidou 2-16 (Compass G-6)	China	Chinese Defense Ministry	25/10/2012
2016-037A	41586	Beidou 2-23 (Compass G-7)	China	Chinese Defense Ministry	12/06/2016
2015-019A	40549	Beidou 3I-1S	China	Chinese Defense Ministry	30/03/2015
2015-053A	40938	Beidou 3I-2S	China	Chinese Defense Ministry	29/09/2015
2015-037A	40748	Beidou 3M-1S	China	Chinese Defense Ministry	25/07/2015
2015-037B	40749	Beidou 3M-2S	China	Chinese Defense Ministry	25/07/2015
2016-006A	41315	Beidou 3M-3S	China	Chinese Defense Ministry	1/02/2016
2010-001A	36287	Beidou G1 (Compass G-1)	China	Chinese Defense Ministry	16/01/2010
2010-024A	36590	Beidou G3 (Compass G-3)	China	Chinese Defense Ministry	2/06/2010
2010-057A	37210	Beidou G4 (Compass G-4)	China	Chinese Defense Ministry	31/10/2010
2012-008A	38091	Beidou G5 (Compass G-11)	China	Chinese Defense Ministry	24/02/2012
2010-036A	36828	Beidou IGSO-1 (Compass G-5)	China	Chinese Defense Ministry	31/07/2010
2010-068A	37256	Beidou IGSO-2 (Compass G-7)	China	Chinese Defense Ministry	17/12/2010
2011-013A	37384	Beidou IGSO-3 (Compass G-8)	China	Chinese Defense Ministry	9/04/2011
2011-038A	37763	Beidou IGSO-4 (Compass G-9)	China	Chinese Defense Ministry	26/07/2011
2011-073A	37948	Beidou IGSO-5 (Compass G-10)	China	Chinese Defense Ministry	1/12/2011
2016-021A	41434	Beidou IGSO-6	China	Chinese Defense Ministry	30/03/2016
2007-011A	31115	Beidou M1 (Compass M1)	China	Chinese Defense Ministry	14/04/2007
2014-050A	40128	Galileo FOC FM1 (Galileo 5)	ESA	European Space Agency	22/08/2014
2016-030A	41549	Galileo FOC FM10 (Galileo 13)	ESA	European Space Agency	24/05/2016
2016-030B	41550	Galileo FOC FM11 (Galileo 14)	ESA	European Space Agency	24/05/2016

2014-050B	40129	Galileo FOC FM2 (Galileo 6)	ESA	European Space Agency	22/08/2014
2015-017A	40544	Galileo FOC FM3 (Galileo 7)	ESA	European Space Agency	27/03/2015
2015-017B	40545	Galileo FOC FM4 (Galileo 8)	ESA	European Space Agency	27/03/2015
2015-045A	40889	Galileo FOC FM5 (Galileo 9)	ESA	European Space Agency	11/09/2015
2015-045B	40890	Galileo FOC FM6 (Galileo 10)	ESA	European Space Agency	11/09/2015
2015-079B	41175	Galileo FOC FM8 (Galileo 11)	ESA	European Space Agency	17/12/2015
2015-079A	41174	Galileo FOC FM9 (Galileo 12)	ESA	European Space Agency	17/12/2015
2011-060B	37847	Galileo IOV-1 FM2	ESA	European Space Agency	21/10/2011
2011-060A	37846	Galileo IOV-1 PFM	ESA	European Space Agency	21/10/2011
2012-055A	38857	Galileo IOV-2 FM3	ESA	European Space Agency	12/10/2012
2012-055B	38858	Galileo IOV-2 FM4	ESA	European Space Agency	12/10/2012
2011-009A	37372	Glonass 701 (Glonass-K, Cosmos 2471)	Russia	Ministry of Defense	26/02/2011
2014-075A	40315	Glonass 702 (Glonass K, Cosmos 2501)	Russia	Ministry of Defense	1/12/2014
2005-050A	28915	Glonass 714 (Cosmos 2419)	Russia	Ministry of Defense	25/12/2005
2006-062C	29672	Glonass 715 (Glonass 35-1, Cosmos 2424)	Russia	Ministry of Defense	25/12/2006
2006-062A	29670	Glonass 716 (Glonass 35-2, Cosmos 2425)	Russia	Ministry of Defense	25/12/2006
2006-062B	29671	Glonass 717 (Glonass 35-3, Cosmos 2426)	Russia	Ministry of Defense	25/12/2006
2007-052B	32276	Glonass 719 (Glonass 36-2, Cosmos 2432)	Russia	Ministry of Defense	26/10/2007
2007-052A	32275	Glonass 720 (Glonass 36-3, Cosmos 2433)	Russia	Ministry of Defense	26/10/2007
2007-065A	32393	Glonass 721 (Glonass 37-1, Cosmos 2435)	Russia	Ministry of Defense	25/12/2007
2007-065C	32395	Glonass 723 (Glonass 37-3, Cosmos 2436)	Russia	Ministry of Defense	25/12/2007
2008-046B	33379	Glonass 725 (Glonass 38-2, Cosmos 2443)	Russia	Ministry of Defense	25/09/2008
2009-070A	36111	Glonass 730 (Glonass 41-1, Cosmos 2456)	Russia	Ministry of Defense	14/12/2009
2010-007A	36400	Glonass 731 (Glonass 42-1, Cosmos 2459)	Russia	Ministry of Defense	1/03/2010
2010-007C	36402	Glonass 732 (Glonass 42-3, Cosmos 2460)	Russia	Ministry of Defense	1/03/2010
2009-070B	36112	Glonass 733 (Glonass 41-2, Cosmos 2457)	Russia	Ministry of Defense	14/12/2009
2009-070C	36113	Glonass 734 (Glonass 41-3, Cosmos 2458)	Russia	Ministry of Defense	14/12/2009

2010-007B	36401	Glonass 735 (Glonass 42-2, Cosmos 2461)	Russia	Ministry of Defense	1/03/2010
2010-041C	37139	Glonass 736 (Glonass 43-1, Cosmos 2464)	Russia	Ministry of Defense	2/09/2010
2010-041B	37138	Glonass 737 (Glonass 43-2, Cosmos 2465)	Russia	Ministry of Defense	2/09/2010
2010-041A	37137	Glonass 738 (Glonass 43-3, Cosmos 2466)	Russia	Ministry of Defense	2/09/2010
2011-055A	37829	Glonass 742 (Glonass-M, Cosmos 2474)	Russia	Ministry of Defense	2/10/2011
2011-064A	37867	Glonass 743 (Glonass 44-2, Cosmos 2476)	Russia	Ministry of Defense	4/11/2011
2011-064B	37868	Glonass 744 (Glonass 44-3, Cosmos 2477)	Russia	Ministry of Defense	4/11/2011
2011-064C	37869	Glonass 745 (Glonass 44-1, Cosmos 2475)	Russia	Ministry of Defense	4/11/2011
2011-071A	37938	Glonass 746 (Glonass-M, Cosmos 2478)	Russia	Ministry of Defense	27/11/2011
2013-019A	39155	Glonass 747 (Glonass-M, Cosmos 2485)	Russia	Ministry of Defense	26/04/2013
2016-008A	41330	Glonass 751 (Glonass-M, Cosmos 2514)	Russia	Ministry of Defense	7/02/2016
2016-032A	41554	Glonass 753 (Glonass-M, Cosmos 2516)	Russia	Ministry of Defense	29/05/2016
2014-012A	39620	Glonass 754 (Glonass-M, Cosmos 2491, RS-46)	Russia	Ministry of Defense	23/03/2014
2014-032A	40001	Glonass 755 (Glonass-M, Cosmos 2500)	Russia	Ministry of Defense	14/06/2014
2010-022A	36585	Navstar GPS IIF-1 (Navstar SVN 62, PRN 25, USA 213)	USA	DoD/US Air Force	28/05/2010
2015-033A	40730	Navstar GPS IIF-10 (Navstar SVN 72, PRN 8, USA 262)	USA	DoD/US Air Force	15/07/2015
2015-062A	41019	Navstar GPS IIF-11 (Navstar SVN 73, PRN 10, USA 265)	USA	DoD/US Air Force	31/10/2015
2016-007A	41328	Navstar GPS IIF-12 (Navstar SVN 70, PRN 32, USA 266)	USA	DoD/US Air Force	5/02/2016
2011-036A	37753	Navstar GPS IIF-2 (Navstar SVN 63, PRN 01, USA 232)	USA	DoD/US Air Force	16/07/2011
2012-053A	38833	Navstar GPS IIF-3 (Navstar SVN 65, PRN 24, USA 239)	USA	DoD/US Air Force	4/10/2012
2013-023A	39166	Navstar GPS IIF-4 (Navstar SVN 66, PRN 27, USA 242)	USA	DoD/US Air Force	15/05/2013
2014-008A	39533	Navstar GPS IIF-5 (Navstar SVN 64, PRN 30, USA 248)	USA	DoD/US Air Force	21/02/2014
2014-026A	39741	Navstar GPS IIF-6 (Navstar SVN 67, PRN 06, USA 251)	USA	DoD/US Air Force	17/05/2014
2014-045A	40105	Navstar GPS IIF-7 (Navstar SVN 68, PRN 09, USA 256)	USA	DoD/US Air Force	2/08/2014
2014-068A	40294	Navstar GPS IIF-8 (Navstar SVN 69, PRN 03, USA 258)	USA	DoD/US Air Force	29/10/2014
2015-013A	40534	Navstar GPS IIF-9 (Navstar SVN 71, PRN 26, USA 260)	USA	DoD/US Air Force	25/03/2015
2003-058A	28129	Navstar GPS IIR-10 (Navstar SVN 47, PRN 22, USA 175)	USA	DoD/US Air Force	21/12/2003

2004-009A	28190	Navstar GPS IIR-11 (Navstar SVN 59, PRN 19, USA 177)	USA	DoD/US Air Force	20/03/2004
2004-023A	28361	Navstar GPS IIR-12 (Navstar SVN 60, PRN 23, USA 178)	USA	DoD/US Air Force	23/06/2004
2004-045A	28474	Navstar GPS IIR-13 (Navstar SVN 61, PRN 02, USA 180)	USA	DoD/US Air Force	6/11/2004
1997-035A	24876	Navstar GPS IIR-2 (Navstar SVN 43, PRN 13, USA 132)	USA	DoD/US Air Force	23/07/1997
1999-055A	25933	Navstar GPS IIR-3 (Navstar SVN 46, PRN 11, USA 145)	USA	DoD/US Air Force	7/10/1999
2000-025A	26360	Navstar GPS IIR-4 (Navstar SVN 51, PRN 20, USA 150)	USA	DoD/US Air Force	11/05/2000
2000-040A	26407	Navstar GPS IIR-5 (Navstar SVN 44, PRN 28, USA 151)	USA	DoD/US Air Force	16/07/2000
2000-071A	26605	Navstar GPS IIR-6 (Navstar SVN 41, PRN 14, USA 154)	USA	DoD/US Air Force	10/11/2000
2001-004A	26690	Navstar GPS IIR-7 (Navstar SVN 54, PRN 18, USA 156)	USA	DoD/US Air Force	30/01/2001
2003-005A	27663	Navstar GPS IIR-8 (Navstar SVN 56, PRN 16, USA 166)	USA	DoD/US Air Force	29/01/2003
2003-010A	27704	Navstar GPS IIR-9 (Navstar SVN 45, PRN 21, USA 168)	USA	DoD/US Air Force	31/03/2003
2005-038A	28874	Navstar GPS IIR-M-1 (Navstar SVN 53, PRN 17, USA 183)	USA	DoD/US Air Force	26/09/2005
2006-042A	29486	Navstar GPS IIR-M-2 (Navstar SVN 52, PRN 31, USA 190)	USA	DoD/US Air Force	25/09/2006
2006-052A	29601	Navstar GPS IIR-M-3 (Navstar SVN 58, PRN 12, USA 192)	USA	DoD/US Air Force	17/11/2006
2007-047A	32260	Navstar GPS IIR-M-4 (Navstar SVN 55, PRN 15, USA 196)	USA	DoD/US Air Force	17/10/2007
2007-062A	32384	Navstar GPS IIR-M-5 (Navstar SVN 57, PRN 29, USA 199)	USA	DoD/US Air Force	20/12/2007
2008-012A	32711	Navstar GPS IIR-M-6 (Navstar SVN 48, PRN 07, USA 201)	USA	DoD/US Air Force	15/03/2008
2009-043A	35752	Navstar GPS IIR-M-8 (Navstar SVN 50, PRN 05, USA 206)	USA	DoD/US Air Force	17/08/2009
2013-034A	39199	IRNSS-1A (Indian Regional Navigation Satellite System)	India	Indian Space Research Organization (ISRO)	1/07/2013
2014-017A	39635	IRNSS-1B (Indian Regional Navigation Satellite System)	India	Indian Space Research Organization (ISRO)	4/04/2014
2014-061A	40269	IRNSS-1C (Indian Regional Navigation Satellite System)	India	Indian Space Research Organization (ISRO)	15/10/2014
2015-018A	40547	IRNSS-1D (Indian Regional Navigation Satellite System)	India	Indian Space Research Organization (ISRO)	30/03/2015
2016-003A	41241	IRNSS-1E (Indian Regional Navigation Satellite System)	India	Indian Space Research Organization (ISRO)	20/01/2016
2016-015A	41384	IRNSS-1F (Indian Regional Navigation Satellite System)	India	Indian Space Research Organization (IRSO)	10/03/2016
2016-027A	41469	IRNSS-1G (Indian Regional Navigation Satellite System)	India	Indian Space Research Organization (IRSO)	28/04/2016
2010-045A	37158	QZS-1 (Quazi-Zenith Satellite System, Michibiki)	Japan	Japan Aerospace Exploration Agency (JAXA)	11/09/2010

Made in the USA
Las Vegas, NV
07 November 2026

57674074R00069